Magic, Witchcraft and the Otherworld

Magic, Witchcraft and the Otherworld

An Anthropology

Susan Greenwood

Oxford • New York

First published in 2000 by
Berg
Editorial offices:
150 Cowley Road, Oxford, OX4 1JJ, UK
838 Broadway, Third Floor, New York, NY 10003-4812, USA

Berg is the imprint of Oxford International Publishers Ltd.

Library of Congress Cataloging-in-Publication Data

A catalogue record for this book is available from the Library of Congress.

British Library Cataloguing-in-Publication Data

A catalogue record for this book is available from the British Library.

ISBN 1 85973 445 6 (Cloth)
1 85973 450 2 (Paper)

Typeset by JS Typesetting, Wellingborough, Northants.
Printed in the United Kingdom by Biddles Ltd, Guildford and King's Lynn.

To Lauren and Adrian

Contents

Preface

This is a study of certain ideas, philosophies, practices and groups within the Western esoteric tradition in the last decade of the twentieth century. I explore, through issues concerning magical identity, gender, and morality, aspects of what magicians term the 'otherworld'. The otherworld is a spiritual domain that is said to co-exist with the ordinary everyday world; it is at one and the same time primordial and also flowing through time, space and within the individual. My analysis develops an approach that stresses the importance of a reflexive and experiential fieldwork, and I begin this study with a short autobiographical introduction. As a child I was given to spending long periods of time in contemplation and reflection on matters spiritual, finding most connection with nature rather than organized religion. In the late 1970s I encountered feminism, and after reading Starhawk's *Dreaming the Dark* (1982) I subsequently embarked upon an exploration of feminist witchcraft. I was attracted to this form of magic by the combination of nature religion, politics and the promise of psycho-spiritual development. Later, as my experience and my questioning increased, I started an undergraduate degree in anthropology and sociology at Goldsmiths College, University of London. A final year research project on women's spirituality resulted in more questions than it answered; and my exploration eventually led to a PhD thesis, the basis of this book.

Consequently, the research on which this book is founded has been part of a personal drive to make sense of the Pagan alternative to more organized religion. I have attempted to portray the realities of this countercultural movement with the eye of an anthropologist, but also with an empathy of spirit which, although critical of a great deal of what I have encountered, has tried to communicate the essence of magic – of what it is like to be deeply involved in the spiritual experience of the otherworld. My role has been as a communicator between two very different worlds: the academic sphere and that of the magical counterculture. It is a position that has not been easy, and it is one that has, at times, been misunderstood by both sides. However, I have been encouraged by many people, and it is largely due to them that this book has reached publication.

I owe a great deal to three people at Goldsmiths College. Firstly and foremostly, I would like to warmly thank Olivia Harris, who accepted the PhD. research proposal and supervised the greater part of the thesis. Her enthusiasm and commitment to the work helped me to gain confidence in myself and with the material, and to ask appropriate questions during some difficult fieldwork situations. During the latter stages of the writing up Pat Caplan also acted as supervisor, and went way beyond academic duty in her reading of final drafts, as well as in her support and encouragement. Brian Morris first introduced me to the anthropological study of religion as an undergraduate in 1986, and more recently, gave freely of philosophical comment on some inspiring walks over the Sussex Downs.

Particular appreciation goes to Annie Keeley, for her friendship over the years; my discussions with her as a co-explorer in some arcane avenues of thought and practice have been invaluable. Ken Rees read chapters as they evolved, and his knowledge of witchcraft has been a valued resource that he has generously shared. Geoffrey Samuel also read chapters and associated papers, and I have much enjoyed the intellectual stimulation of subsequent discussions, particularly on the other-world. I am also grateful to Roger Greenwood, who has been a constant source of practical help and support in technical matters. I would like to thank all the practitioners of magic who allowed me to share their rituals and gave so generously of their time, hospitality and patience in answering my questions.

Many other people have given assistance in various ways, and I wish to thank in particular: Les Back, Brian Bates, Eileen Barker, Barbara Bender, Chris Boyes, Nickie Charles, Peter Clarke, Simon Coleman, Yvonne Eadie, Celia Glass, Victoria Goddard, Charlotte Hardman, Graham Harvey, Paul Heelas, Jos Hincks, John Hinnells, Felicia Hughes-Freeland, Chris Knight, Jean La Fontaine, Nici Nelson, Stephen Nugent, Joanna Overing, Peter Pels, Liz Puttick, Yvonne Spence, Robin Vincent, Diana Stone and Michael York.

The research was funded by grants from the University of London Central Research Fund and the Economic and Social Research Council UK, and I am appreciative of their financial assistance.

Finally, I would like to thank Adrian and Lauren for putting up with a mother who has, for most of their lives, done strange things – such as taking off to the woods in the middle of the night – in the name of anthropological research. They have also shown great tolerance of the demands of writing, and it is to them that I dedicate this work.

–1–

Introduction

Anthropology has traditionally been concerned with the translation of cultures usually of non-Western small-scale or developing societies; but this book is an anthropological study of a Western magico-spiritual counter-culture commonly termed Paganism. Paganism is an umbrella term for a number of diverse groups and practices, and ranges from various forms of high magic (sometimes called ceremonial magic or 'Western mysteries'), through witchcraft or wicca, druidry, and the Northern magical traditions of the Scandinavian and Germanic peoples, to chaos magick.[1] These disparate groups have varying mythologies and cosmologies; but all share a common uniting belief in communication with an 'otherworld' – a realm of deities, spirits or other beings experienced in an alternative state of consciousness. The otherworld is viewed as part of a holistic totality co-existent with ordinary, everyday reality; it is seen to be a source of sacred power. Contact and communication with the otherworld is usually conducted through special rituals, a process that is seen to bring transformations both to the individual and to the wider cosmos. Connection with the otherworld is thus central to the magical practice and experience of Paganism.

Engaging in magical rituals designed to facilitate contact with the otherworld affects magicians' notions of identity, gender and morality. These areas of magical life illustrate how central the experience of the otherworld is to magical thought; they also show decisive differences not only between different magical philosophies but also between the magical world and mainstream society. Since the Enlightenment, Western cultures have been associated with reason and rationality, which are in turn connected to notions of masculinity (Seidler 1989). Magicians argue that there are multiple ways of viewing the cosmos, and that the rational, objective realm is not the only reality. Magical rituals in contrast provide a space devoted to the forces of 'unreason' – whereby everything associated with emotion, intuition, and femininity is valued. It is a striking feature of magical practices that often women or 'the feminine' are central, in contrast to orthodox world religions, where women have largely been marginalized. This is because most contemporary magical ideologies are based on the predominant stereotypical view of gender difference: women are more intuitive, men more rational. Being intuitive is seen to be an essential characteristic of a good magician, who must be in touch with the forces of the otherworld; and women are often thought to have an innate advantage. In some practices, such as witchcraft, women act as initiators of men into magic. In addition, magical energy is often likened to the two poles of a battery, one of which must be negative, the other positive for the current to flow. Sexual stereotyping positions women and men as opposing poles in the energy field. Rituals form a theatrical space apart from the ordinary world, in which the body is seen to be the locus of the forces that are personalized as spirits or deities of the otherworld. For magicians, ritual is a space of resistance to the rationalism of the wider culture. Rituals are viewed as a space where a magician gains contact with the otherworld, a special 'place between the worlds', where magical transformations are said to occur.

Magic, to the outsider's eye, is concerned with mystery and beguilement; it is the stuff of enchantment, popularized and synthesized by the films of Disney and synonymous with fantasy and dreams. Real-life magicians take on some of the popular images of magic, and many weave them into magical personas that are associated with an otherworldly power. However, underneath the glitter and glamour of magic there is another more serious process of psychospiritual transformation. Magicians' identities are formed from their relationship with the otherworld, and part of the practice of magic as a spiritual path requires learning how to channel the forces of the cosmos. But before this can happen much magical work involves healing the magician from the effects of living in the ordinary world; and the parallels with shamanism are explicit. It is only when the magician can 'balance the forces' that the work of bringing power through from the otherworld begins.

Morality in magical practices is an important issue. For magicians, good and evil are joined in a dialectical process that creates a greater cosmic unity: there is no absolute separation between good and evil. The 'occult' means that which is kept secret, the esoteric, the mysterious, that beyond the range of ordinary knowledge. In Christian terms, it is often thought of as sinister, and associated with the Devil as an opposing moral force to the goodness of God. The term 'occult' is loose and all-embracing, and can refer to a variety of practices ranging from Spiritualism to alchemy or astrology. It is often used by outsiders in a derogatory way to describe that which is seen to be heretical to orthodox beliefs. In particular, witchcraft has been associated with evil, darkness, illness, misfortune and death. Christianity constructed the image of the witch as an associate of the Devil in an evil partnership that threatens an ordered and good society. In this way, notions of witchcraft may be seen to embody a society's negative values, as in the case of the witches' Sabbath, which is associated with gatherings in the darkness, female sexuality and animal metamorphosis. This is in contrast to the Christian emphasis on light, controlled sexuality or chastity and definite boundaries between animal and human (Ginzburg 1992). Morality in magic is thus effectively shaped by the Christian religion of the mainstream society; and I show how magical practices either incorporate elements of Christianity or rebel against it in an effort to determine an internal sense of morality that is in harmony with natural 'cosmic truths'.

Often anthropological and sociological analyses of magic fail to take account of magicians' interactions with the otherworld. This is due to the fact that the social sciences have had a problematic relationship with magic, largely as a result of the close association of rationality with a certain view of science. By failing to attach sufficient importance to the otherworld, these analyses miss what Pagans see as the essence of magic: otherworldly experience. By focusing on central issues concerning magical identity, gender, and morality the value of a different perspective that incorporates ideas and experiences of the otherworld becomes clear. This work has implications for the study of the social sciences as a whole, and engages with perennial philosophical and anthropological debates concerning what rationality is and what constitutes a scientifically valid knowledge of reality. This study is part of a growing corpus of work that questions ethnographic epistemologies and the construction of knowledge through cultural interpretation (Favret-Saada 1980; Marcus and Fischer 1986; Stoller and Olkes 1987; Tedlock 1991; Young and Goulet 1994; Cohen and Rapport 1995). However, before discussing these issues in more detail I will introduce the magical subculture in London.

The Magical Subculture in London

'Talking Stick' is an informal discussion forum that is held twice a month when a large group of a hundred or so magicians gather in the upstairs meeting room of a central London pub near the British Museum. Nearby stands the empty Centrepoint building, which occupies a site that, according to magical lore, was cursed by the notorious magician Aleister Crowley. On one occasion when I attended one such meeting the atmosphere in the room was hot and smoky, but also charged with expectation, for the speaker was a well-known personality in the magical scene, and most were interested in what she had to say. Down below the traffic roared; but the outside world seemed very distant on this hot May evening. Recent forum talks had included 'Jung and the Uncanny (Making Coincidences Meaningful)', 'Does Magic Work?', 'Witchcraft and Tradition in The Basque Country', 'Renaissance Magic', and a 'Meet The Groups' session with representatives of different magical groups.

Magical practitioners of all persuasions gathered in small groups, exchanging information and chatting. Most were in their twenties and thirties, white, middle-class and not really out of the ordinary at first glance. On closer inspection pentangles and talismans, in the form of rings or other decorative jewellery, became more apparent. There was also a preponderance of black clothing, especially black leather. Leaflets were circulated: a member of a Pagan group was having an open house and 'workshops, story-telling, a discussion forum, an attunement meditation and short closing ritual' were on offer. Information on magical moots in another area of London, including 'Themes of Midsummer', and 'Sacred Plants and Natural Magic', were also available. For those with an interest in Egyptian magic, a Midsummer Ritual of 'The Rebirth of Amun Ra' was advertised. This was to be led by an archaeologist specializing in Egyptology and his partner, who were also high priest and priestess of a witchcraft coven. 'Together they will provide an opportunity to understand and participate in the voyage of the Sun god, Ra, across the horizon. Using visualisation and magickal technique, we will experience the power of solar transformation.' Various Pagan journals and newspapers, including *Psychic News,* were on sale from a central table, as were various books relevant to the talk.

How does the newcomer into the occult choose from such a bewildering array of paths to 'spiritual truth'? There are perhaps hundreds of open organizations, ranging from Charlton House Astrology Group through The Fellowship of Isis (created to promote closer communication with the Goddess), Cheiron (a group for elderly, disabled and invalid occultists), the Eagle's Wing Centre for Contemporary Shamanism, Hoblink (a networking group for homosexual/bisexual Pagans), and the Order of Bards Ovates and Druids, to the Theosophical Society and The Vampyre Society. There are experts who will teach various aspects of Kabbalah[2] in weekend courses with such titles as Kabbalistic Astrology and 'The Tree of Life'. Alternatively, it is possible to join a 'Wicca Study Group' to discover, amongst other things, the Wiccan Circle and the Elements, The Wheel of the Year, The Goddess and the God, Making Magic and the Meaning of Initiation.

Most of these magicians call themselves Pagans or neo-Pagans,[3] and 'Paganism' has become the descriptive term for a mulitiplicity of diverse practices. Some, usually those in the northern traditions, prefer to be called 'Heathen' (Harvey 1996:49); but Paganism has become the descriptive label for this type of magico-spiritual practice. However, this is not unproblematic, because Paganism has a rather vague umbrella usage and refers to a group of magical practices that are often seen to pre-date Christianity. The word 'pagan' is derived from the Latin *pagus* meaning 'rural', 'from the countryside', and has often been used to designate the 'other' from Christianity (Jones and Pennick 1995). Contemporary self-designated Pagans use the term broadly as 'one who honours nature'. A current definition of Paganism is as 'a Nature-venerating religion which endeavours to set human life in harmony with the great cycles embodied in the rhythms of the seasons' (Jones and Pennick 1995). I argue that far from being prior to Christianity, contemporary Pagan practices are either a Christian esoteric interpretation of magical beliefs, or have developed specifically in opposition to Christianity. A study of history shows that many current magical practices stem from revivals of magic in the Hermetism of the Renaissance and, more recently, in the nineteenth-century magical organization the Hermetic Order of the Golden Dawn. In other words, contemporary Paganism is based more on Hermetism rather than any indigenous nature religion. However, the term Paganism has gained widespread currency, and is used to describe the increasing number of people who practise what they define as an ancient nature spirituality. Paganism is part of a wider Western trend towards a de-traditionalized and internal spiritual experience rather than a received 'religion' external to the individual and imposed from the outside.

Consequently, many magicians practise magic as part of a critique of Christianity, which most described to me as 'dualistic' in the way that it splits the body off from an externalized spirit. Many magicians construct their philosophies in terms of being oppressed by Christianity, as a verse of 'The Burning Times', a song composed by the popular Pagan rock band Incubus Succubus,[4] states:

> They came to bring the 'good news',
> to burn Witches, Pagans, Jews.
> They said they were the shepherd's sheep,
> they whipped old women through the street.
> Then the turning of the tide,
> from the truth they could not hide.
> Now the darkest age is past,
> the Goddess has returned at last!

However, Paganism also encompasses groups that are broadly Christian. Indeed, some high magicians are committed Christians, such as Dion Fortune (1987a), who expounded an esoteric version of Christianity. Although many Pagans see their practice as antithetical to Christianity, some magical practices are directly shaped by Protestant notions of the individual and his or her relation to God. The case of the infamous magician Aleister Crowley, who rebelled against a strict Protestant upbringing and developed the anarchistic religion of Thelema as a direct response, is a good example. It is thus clear that magical practices have a complex relationship with Christianity, and are often not so different from Christianity as they claim to be.

The exact number of practitioners of magic is notoriously difficult to calculate. Most are anarchistic and have an aversion to filling in census forms, and the very nature of magical practice is couched in secrecy fomented by mythologies (of the 'Burning Times' for example) and actual discrimination and abuse (one witch I spoke to said that he had had bricks thrown in though his window and his door smashed down). Witchcraft in particular is decentralized, and each coven is autonomous.[5] Figures of several thousand in England have been suggested (Luhrmann 1989). York refers to a census organized by the occult supplier 'The Sorcerer's Apprentice' in which Chris Bray, the proprietor, claims that 'at the present day we have a conservative estimated population of over 250,000 Witches/Pagans throughout the UK and many more hundreds of thousands of people with a serious interest in Astrology, Alternative Healing Techniques and Psychic Powers' (The Occult Census 1989. Leeds. Quoted in York, 1995:143). The Pagan Federation states that its membership figures in 1992 were 1,200 full members, with approximately the same number of associate members.[6] In 1994 it claimed that it had 'grown fivefold in the last two years' (Talking Stick Magickal conference 12 February 1994). In 1997 numbers at the annual conference, which was held at a much larger venue, doubled from 850 of the previous year to 1,500 ticket sales (with stall-holders and tickets for the night concert this approached the figure of 2,000). Paganism is without doubt an increasingly popular form of spirituality, and the numbers of those who call themselves Pagan will undoubtedly rise in the years to come.

Attendance at the London events such as the annual Pagan Federation or Fellowship of Isis Conferences typically drew around four hundred participants when I was conducting fieldwork in the early 1990s. A popular discussion forum, such as Talking Stick, would attract, on a weekday evening, at least a hundred magicians. New Age and Pagan bookstores, which also sell magical paraphernalia (candles, incense, figurines, etc.), are increasingly becoming a standard feature in major British cities. From my own involvement I can say that in October 1993 I was aware of at least eleven witchcraft covens (consisting of four or five members up to the traditional maximum of thirteen) in London, and there were probably many times that number that I was unaware of. High magic groups are usually started through correspondence courses, which take students from all over the country. There are local lodges, and students usually use their nearest one when they become involved in group ritual work. Two London lodges were known to me, and I heard that the occult school of which I was an apprentice had ten lodges in other countries. These impressionistic figures do not take into account solitary practitioners or other practices and traditions within the subculture as a whole, and any guess at numbers inevitably remains an approximation.

My focus in this work is primarily on the distinctions and similarities between high magic on the one hand, and forms of witchcraft on the other. This comparison reveals certain tensions and differences in the ways that contemporary magicians distinguish their practices. Since the Renaissance, high magic has been concerned with drawing down forces and energies from the heavens. Christian-influenced high magic mythology relies on a classic narrative: humans master their lower selves and baser natures in the spiritual pursuit of their true identity within the Light and legitimacy of the Ultimate Being. The aim of high magic is wholeness and unity with divinity. Wholeness in high magic is commonly explained with reference to the Judaeo-Christian Fall – the separation of humanity from godhead and its subsequent reunion.

Some practices of high magic, following the ideas of Aleister Crowley, are explicitly anti-Christian, and focus on 'liberating' energy to develop the self rather than on notions of spiritual union with divinity. Contemporary witchcraft is also anti-Christian, and was created in the 1940s by a retired civil servant called Gerald Gardner (Kelly 1991; Hutton 1991). It is claimed to be a re-working of ancient magical practices – the 'Old Religion' – and part of an ancient fertility religion (Gardner 1954); it is also said to be a nature religion. I argue that modern witchcraft is not a form of 'low magic' associated with ordinary peasant folk and country lore, but rather is a development of the Renaissance high magic tradition, and can thus be associated with the learned and élite rather than ordinary village folk. This view conflicts with contemporary witches' own mythology of 'the Burning Times' – the European witch hunts of 1400–1750 – which represent a powerful metaphor for their oppression by the dominant rationalist culture, symbolized by Christianity. In other words, the fact that witchcraft can be demonstrated to have derived from high magic challenges the very essence of its ideology as an ancient pre-Christian pagan tradition. Gardner's formulation of witchcraft practice concerned the channelling and utilization of the sexual energy of coven members. Later versions – Alexandrian witchcraft, which was developed by Alex Sanders in the 1960s, and incorporated more high magic elements (often referred to by witches as 'high church'), and feminist witchcraft, which was formed to counter the male bias in mainstream orthodox religions (see Chapter 4) – have also been influential on the subculture as a whole.

Perhaps the majority of Western magicians are brought up within a culture which has implicit and explicit Christian values. Many people search for alternatives to Christianity. This may involve experiencing and experimenting with new forms of spirituality. Steven, one of my informants, explained the route on which his spiritual search had taken him. He was in his early forties, had come from 'basically a Christian family', and was typical of many Pagans in the way that he had explored and practised both witchcraft and high magic. When he was young, Steven heard that witches worked in a secluded spot in a wood at Halloween. He and his friends decided to visit the wood at the appropriate time to see if the witches were there. Amazed actually to find them, he told me that he felt all the fears and superstitions about witches that he had learnt through his Christian upbringing. Terrified, he and his friends ran off, but Steven fell and was caught by the witches. They brought him into the centre of their circle and talked to him. He was very frightened, but the witches explained that they were 'not what he thought'. Steven told me that he had always been naturally curious, and that he took the opportunity to ask them questions about what they were doing.

This experience prompted Steven to think about religion, and he started searching 'in a spiritual sense'. He looked at Buddhism; but although he was very attracted to it, he felt that it was not right for Europeans. He eventually became a priest in the Isian religion[7] (a form of high magic) in 1984 after he had 'found' Isis in a cinema at Newmarket, whilst watching Ursula Andress playing the part of Isis.

The name triggered something in his unconscious mind, and he set out to research the Egyptian pantheon. Some time later, when he was wandering around the West Country feeling depressed, he met a group of people in a hostelry in Bristol and got into conversation with them. They turned out to be members of an Order of the Priests of Isis, and Steven was eventually 'led to their temple'. He was created a priest to learn, and was sent out to spread the word by 'informal chats'. Steven met working witch groups and was also finally initiated as an Alexandrian witch. 'Using his own interpretation' he was instructed to teach as a 'continual learning process'. He was a priest in an Alexandrian-based coven until 1990; but he now distances himself from Alexandrian witchcraft, preferring to practise what he calls traditional witchcraft or the 'Craft of the Wise' – the practice of the historical cunning man.[8] He views himself as a seer and a healer, and people come to him for alternative medicine, aromatherapy and divination.The rituals that Steven performs now are a mixture of witchcraft and Isian temple workings.

Steven practises both high magic and witchcraft. Many magicians move from one tradition to the other, or more commonly, as Steven has done, incorporate both aspects into their magical workings. Magical practices lend themselves to an almost constant 'mix and match', and, while separate philosophies and cosmologies define certain differences, actual magical practice is not a neat and tidy affair.

Paganism and the New Age

How do magical practices fit into the general religious counterculture? The period following the Second World War has seen a burst of religious creativity that has been noted by a number of authors on new religious movements (NRMs) (for example Nelson 1987; Robbins 1991). In such sociological analyses,[9] counter-cultural movements or cults – such as those within the occult subculture – have been defined as 'religious movements which draw their inspiration from elsewhere than the primary religion of the culture, and which are not schismatic movements in the same sense as sects, whose concern is with preserving a purer form of the traditional faith' (Glock and Stark 1965:245). One feature of cults is their supposed ephemeral nature. However, the historian of religion Gordon Melton is critical of this notion, and claims that cults are not rootless or transitory but products of the massive diffusion of the world's religions globally (1993:97). Thus 'new' religions of the 1970s are not actually new, but should be understood as a continuation of historical events into contemporary magical ideas.

This position runs counter to much work in the sociology of religion that focuses exclusively on the present impact of NRMs. Beckford (1985), for example, argues that new religious movements may not be of recent origin, but that their collective impact is novel. This position is also taken by Michael York, who, in his study of the New Age and Neo-paganism, argues that a case for historical antecedents can be made, but that it is the social impact of new religious movements in contemporary society that exemplifies their 'newness'. He is interested in the perception of them as 'sudden and unprecedented developments' (1995:7). The value of this approach is the focus on the immediacy of their appeal at certain times and in particular social conditions. Beckford (1984) argues that the appeal of movements that stress holistic conceptions of self caters to people fragmented into diverse functionally-specific social roles. However, the problem with this approach is that the field of study becomes vague, and very different practices tend to merge into one another. York claims that the New Age and Neo-paganism are both manifestations of the Western occult tradition, but that increasingly Paganism is becoming bound up with 'Eastern mysticism/human potential and the theosophical-occult/ spiritualist-psychic/new thought metaphysics mix' as one of the main constituents within the New Age (1995:99).

The contemporary practice of magic has been facilitated by the New Age movement, which according to Gordon Melton (1986) can best be dated from 1971.[10] The central vision of the New Age is one of radical mystical transformation, a view of a new world that transcends the limitations of any particular culture, religion or political system. This may involve an awakening to new realities, such as the discovery of psychic abilities, or physical or psychological healing and the emergence of a new vision of society, based on an enlarged concept of human potential, which combines some of the latest scientific discoveries with 'ancient' spiritual practices. The New Age has opened up the sphere of 'mysticism', of which magical practices are a part. William Bloom, a leading exponent of the New Age, writes that all the different dynamics of the New Age movement 'thread together to communicate the same message'. This message is that there is 'an invisible and inner dimension to all life – cellular, human and cosmic. The most exciting work in the world is to explore this inner reality' (1991:xvi).

The British occult subculture shares the holistic philosophy and many of the broad ideals of the New Age movement. York notes a number of similarities and differences between the New Age and Neo-pagan movements (although they are not easily distinguishable). Similarities include a concern with healing and a holistic emphasis. York maintains that both movements may be seen in part as developments of and outgrowths from the human potential movement, and have the following characteristics: eco-humanism – the belief in the intrinsic divinity of the individual; a focus on personal growth; a reluctance to over-institutionalize; the recognition of the need for a spiritual idiom in feminine terms; a sense of animism (the world is seen to be alive); an emphasis on the non-rational; a belief in reincarnation. Both movements are also characterized by short-lived groups; and finally, both elect antagonism from Christianity.

York suggests that differences cluster around a series of important notions: divinity, tradition and the otherworld. The New Age pursues transcendent metaphysical reality, while Neo-paganism seeks an immanent locus of deity; in terms of tradition the New Age is a new and distinct religion (with an 'almost millennial expectation of quantum change'), whereas Neo-paganism stresses its links to the past; in terms of the otherworld the New Age inclines towards a hierarchical understanding of the supernatural and has a more passive approach to it (for example, meditation or memories from previous incarnations) compared with the more democratic model of Neo-paganism, which is also more active (particularly in relation to ritual). Another difference lies in relations between this world and the otherworld: the New Age tends to de-emphasize the material and emphasize the spiritual, while Neo-paganism is seen as 'perhaps more balanced'; furthermore, the New Age stresses 'White Light', whereas Neo-paganism incorporates both the light and the dark (1995:162).

Both the New Age and contemporary magical practices reflect cultural trans-formations in the West, involving a shift from what some have termed 'paternalistic' authority to autonomous authority – a search within the self rather than acceptance of having meaning conferred externally (Anthony and Ecker 1987:9; Kohn 1991). In his study of the New Age movement, Paul Heelas has coined the term 'self-spirituality' to express this new source of meaning, which involves a shift from what is perceived to be a contaminated mode of being (due to socialization) to a realm of authentic nature. This inner realm may be experienced among other things as 'God', 'Goddess', or most frequently as 'inner spirituality' (1996:19). Heelas compares tradition as a source of authority with the 'de-traditionalization' of the New Age. 'The traditional' is defined as that which speaks with the voice of 'external' and established authority. De-traditionalization is the process whereby 'such voices lose their say'. They come to be replaced by the authority of the individual. New Age practices aim to develop the authority of the 'voice within', as Heelas comments: 'The dictates of all supra-Self 'others', which help construct the ego, should be rejected in favour of that authority which comes from the Self itself.' The New Age, according to Heelas, rejects 'beliefs' in favour of what Tipton (1982) terms an 'expressive ethic', and what lies within provides the sole source of 'genuine ethicality' and no one can exercise authority over anyone else (1993:109).

Like the New Age, the magical subculture consists of practices that can also be described by this expressive ethic: both are concerned with inner transformation. The occult subculture has developed from the same roots as the New Age; but, as York notes, it does have its own quite specific ritual and ideological features. While the British occult subculture may have 'postmodern' tendencies (towards eclecti-cism and diffuseness – reaching its extreme expression in chaos magick), in general it is not postmodern. It is more plausibly a reaction to postmodernity, whereby magicians search for a true identity beneath the superficiality of contemporary culture. On the other hand, some, such as Gellner, argue that 'fundamentalism', i.e a search for absolute certainty through religion, is also 'postmodern' in the sense of being the other side of the coin from scepticism and fragmentation (1992b).

Magic From the Inside

Magic and the study of witchcraft is a classic anthropological concern; but research in a complex society – a non-classical field – brings specific problems. Magicians in Britain are usually highly educated and creative, often holding responsible and demanding jobs. Contemporary magic is a literary-based practice, and a profusion of books and magazines have been written on all aspects of magical philosophy and technique; magicians read voraciously, write a great deal and conduct debates over magic issues through newsletters and magazines. Various magical experts run courses, which often have a strong intellectual as well as experiential compo- nent. It is quite likely, therefore, that magical practitioners will have read classic anthropological texts, and this has offered interesting challenges to me as an anthropologist. Like any anthropologist, I have had to battle with emic and etic interpretations of data; however, while magicians do not share the etic approach, they do incorporate many anthropological concepts in their emic understanding of their practices, and so my fieldwork material is constructed through a reflexive dialogue with aspects of anthropology itself.

It has traditionally been the role of the anthropologist to describe and make intelligible, for a largely Western audience, the strange beliefs and experiences of informants who are largely non-Western. Above all, the anthropologist's role has frequently been one of empathetic but detached observer. Instead of investigating the 'truth value' of informants' experiences of 'seeing spirits' or 'travelling to other realms' anthropologists have used functionalist, structuralist or symbolic models to explain informants' experiences. Alternatively, they have analysed informants' accounts as 'texts' to be analysed in terms of meaning. The emic reality of inform- ants has been treated as interesting and even reasonable (given the premises upon which it rests), but not as a serious alternative to Western scientific views of reality (Young and Goulet 1994). Anthropological methodology has reflected great changes in thinking within the discipline in the last fifteen years or so. George Marcus and Michael Fischer argue that the development of a distinctive anthropological cultural critique is inherently linked to ethnographic experimentation. A distinctive feature of this experimentation is the sophisticated reflection by the anthropologist on herself and her society that is engendered by describing an alien culture. Marcus and Fischer point out that classical ethnography sought to romanticize and make vivid the situation of the fieldworker in a demonstration of how exotic customs made sense in their own contexts. By contrast, many contemporary ethnographies make the situation of the fieldworker problematic and even disturbing for the reader by exploring the philosophical and political problems involved in cultural translation[11] (1986:48). They claim that what they term modernist texts – which problematize the insider/outsider position of the anthropologist – constitute a radical shift in perspective on what ethnographies are about and how they are written. Two characteristics of such texts are that, firstly, the conventional concepts of culture are questioned – they do not rely on a conventional notion of a shared cultural system on which to build. Secondly, they question the coherence of culture. One particular method of conducting research within this framework is a dialogue of reflection upon the experience of another culture. This may involve the anthropologist's experience of culturally-interpreted altered states of consciousness – what Marcus and Fischer term 'sorcerer's apprentice' ethnographies (1986:67–73).

More recently some anthropologists have advocated a more experiential approach: for example, Anthony Cohen and Nigel Rapport in the introduction to *Questions of Consciousness* state that 'we are not doing the job if we do not apply insights about our own consciousness to the problem of interpreting others'. Cohen and Rapport argue that the anthropologist uses his or her own consciousness as a device to imagine and then portray the consciousness of others. The emic consciousnesses should be more accessible because of the redundancy of the old orthodoxy that called for the strict difference between the anthropologist and the anthropologized: 'Modern anthropology is conducted as a dialogue in which, so far as possible, the differentials of power are removed, minimized or neutralized' (1995:9–10).

This study involves a dialogue between academic anthropological discourses and an internal examination of the process of magical practice. As an anthropologist I have sought to examine the process of engaging with the otherworld. My approach in this work is to examine magic from 'the inside'. My aim is a self-conscious reflexive engagement with magical practices, and my contention is that if an anthropologist wants to examine 'magic' then she or he must directly experience the otherworld. Consequently, some of my anthropological data will consist of my direct experience during my magical training, and are thus similar to the work of both Jeanne Favret-Saada (1980) and Paul Stoller (1987). Both these anthropologists became intimately engaged with magical practices. Favret-Saada, in a study of witchcraft in the Bocage region of Normandy, found herself drawn into a strategy of words that had the power to attach themselves to the body or belongings of a listener, who would, in time, become bewitched. She reports that in the Bocage nobody ever talked about witchcraft to gain knowledge – witchcraft was about power. Before she had a chance to utter a single word she was involved in the same power network. A subject was either a witch, an unwitcher or a victim. Favret-Saada explains how interlocutors immediately tried to identify a strategy by estimating the force of words to establish whether a person was a friend or foe, to be bought or destroyed. In this deadly circuit of words a conventional ethnographer's role was impossible, and Favret-Saada had to assume the role of assistant to an 'unwitcher'. This changed her perception of the conventional objective stance of the ethnographer – to differentiate between ordinary language and witchcraft discourse it was necessary for her to engage fully in the discourse.

In a similar manner Paul Stoller was deeply involved in magic when he became an apprentice to Songhay sorcerers (*sorkos*) – a bird sitting on his head was the sign that he had been chosen. Stoller reveals that he could not refuse the opportunity to learn magic from the inside, nor could he resist a 'private desire to become a more powerful person' (1987:27). According to Stoller, the Songhay consider themselves powerless pawns in the scheme of a powerful and deadly universe, controlled by God, his messenger Ndebbi, angels, ancestors and spirits. Incantations come from the heavens, passed down from Ndebbi to the ancestors, and enter the minds of the *sorko*. Through learning texts and incantations the *sorko* may become a powerful healer – he can transform himself into various objects so that an enemy will not see or capture him (1987:88).

This work includes material from my training both as a high magician and as a witch, and I will argue that this deliberately participatory approach is essential to an understanding of contemporary Western magicians' otherworlds, and as such is a valuable tool of research and should not be contrasted with 'scientific truth' or seen to threaten the anthropologist's objectivity. It does, however, raise important theoretical implications for the anthropological study of magic, and confronts the limitations of the Western philosophic tradition (Stoller and Olkes 1987). My aim, like Jeanne Favret-Saada's, is to take magic and the otherworld seriously as a subject of study in its own right and not as a 'metaphor' or 'conceptual device' (cf. Edith Turner, 1992, 1994). Magical energies and forces make up the Western magicians' holistic world. Magic, however, is a force that cannot be empirically identified or measured. If magic and the magicians' otherworlds are seen as irrational or are identified as solely due to individual psychology or as figments of the imagination, or even if – as in Evans-Pritchard's study of the Azande – they are rational in themselves for ordering action and social life, but in the final analysis are inferior to science – then this devalues the reality of magic for the practitioners themselves. Magic is either prior (Tylor, Frazer) or inferior (Evans-Pritchard). Favret-Saada points out in her ethnography of witchcraft in the Bocage region of Normandy that social scientists have the same response to the term 'magic': it is used as a widely-held negative definition 'relating the native to his "otherness"' (1980:198). At the heart of the issue is the co-existence of two incompatible physical theories, one of which is seen to be superior to the other because, since the Enlightenment, the magical has been progressively devalued by the rise of rationalism.

Engaging in fieldwork on the magical subculture meant that I had to deal with the perennial anthropological dilemma of learning how far it is possible to be simultaneously subjective and objective – being an 'insider' and 'outsider'. There were two dimensions of the insider/outsider dichotomy for me as fieldworker to negotiate: firstly, the social structural location of the magical subculture – esoteric groups engage in what is seen to be secret arcane knowledge, and are thus set apart, or set themselves apart, from more mainstream society; and secondly, the physical aspect of engaging with my subconscious as a part of the research. Engaging in a reflexive study of 'the essence of magic' involved an opening up of my subjective self – as 'internal' emotional awareness – beyond that usually required of an anthropologist conducting fieldwork. Extracts from reports from the field sent to my academic supervisor during the period of my fieldwork reveal the complexity of experience that this engendered. The first excerpt concerns a high magic seminar, and demonstrates the contrast between the everyday world and the evolutionary spiritual mythology that shapes the practice and worldview of this particular group of magicians (see Chapter 3):

> As it was such a lovely day we had the first part of the seminar in the back-yard, which was beautifully planted with tubs of wisteria and clematis which entwined over a trellis to make a delightful secluded spot. It was here that I had my usual schizophrenic experience of being on the inside and the outside. I was listening to the Adept going on about the breeding rituals and breeding programmes of the Atlanteans who were seeking, through evolution, 'more perfect forms to house the force of the Higher Self', and at the same time I could hear the everyday noises of the everyday world, the 'less evolved' carrying out their 'less evolved' business – someone was kicking a beer can down the street! (Report from the field no 21, 1 June 1993).

Another example, this time from a report on a witchcraft ritual:

> [The magical group] sat in a beer garden and talked about magical identities. Eventually the others arrived (two women and a man) and we looked over various xeroxed photo-copies of rituals. Usually it is decided well in advance which ritual is going to be done – often rituals are specially written, handed out in advance and lines learnt by heart – but this time the person responsible for organizing the ritual had 'gone missing' and left no message as to her whereabouts. It was decided to do a particular ritual which had been written by someone (they couldn't remember who) at another ritual and passed around. We talked tactics – who was going to be priest and priestess and who was going to say what. I had my usual feeling of being in two worlds at once, the one of the pub beer garden, of talking, laughing groups of people sitting in the evening sun or playing billiards, the other of the secret 'otherworld', the world of the dark wood, the mysterious forces that we would be evoking in the dark away from the lights and the ordinary. We eventually finished our drinks and walked down to the cars where we picked up our ritual paraphernalia (lanterns, wand, chalice, a bag of rose petals, spring water etc.). We started our trek in the dusk to what seemed like the very centre of the wood, the longer we walked the quieter it got (Report from the field no 22, 1 July 1993).

These extracts from my reports reveal some of the difficulties involved in conducting fieldwork in such situations. My feelings of schizophrenia were induced by the awareness that I, as fieldworker, was always in two worlds. Even though I participated fully in magical experiences I could never totally become a magician. I could never switch off the constant observation and questions in the back of my mind, and I soon learnt that it was not acceptable to question too much – emotions, rather than rational thought, were given priority. I also had difficulties with divided loyalties in the sense that the gaining of information – the casual chatting to people rather than formal interviews – for research occasionally felt predatory. Even though I always tried to make sure that whoever I was talking to knew that I was an anthropologist conducting research, I still felt, at times, as though I was prying and 'scavenging' for data. Taking people's subjective accounts of their lives and writing about them in a formal academic manner sometimes made me feel uncomfortable because I was exposing intimate experiences to a wider, and, in many cases, sceptical, if not hostile, audience. Despite these difficulties, I carried on with the research because I was in sympathy with the broad principles of Paganism and thought that my research could increase academic understanding of magical practices.

The practice of magic deeply engaged me, and I had some very intense emotional experiences resulting from my research. In my very first report from the field to my academic supervisor I was concerned with the aspect of using myself – and my subjective experience – as part of the study:

> I think it is important to undergo the experience of opening oneself up to the esoteric practice, to be able to write about that experience from the inside, but a question that has been troubling me is how far can I take this before it becomes a personal indulgence – reflexive anthropology taken to its extreme? My response is that I have to feel the experience to know (feel) what people are talking about but to keep this in balance all the time with a critical 'anthropological eye', always being aware of the focus of the research and its political implications (1 October 1991).

In the early stages of the research I started experimenting with 'otherworldly forces':

> I have been in a very strange space these last couple of weeks, mainly due to my unconscious and conscious identification with what I've been reading on 'female' shamanism . . . I decided to engage with the Goddess as a shamanistic experience. I have experienced the most intense feeling of physical disintegration – coming face to face with death in my imagination . . . I have established a 'relationship' with what I choose reservedly to call 'the Goddess'. This has been a very personal experience and one I am uncertain how to handle with regard to the research data, but it is an experience which I feel is very relevant to the work (letter to my supervisor 23 November 1991).

The shamanic approach is proving useful in understanding the subjective experience of magic. I have been experimenting in this area by 'opening myself up to the experience' of the Goddess and the God in my imagination. It certainly works as a way of ordering and understanding life (as EP said about the Azande) (Report from the field no 4, 1 December 1991).

My time became progressively more focused on subconscious exploration when I joined a high magic school:

I have heard from my [occult] supervisor, and my first lessons have involved meditation exercises and visualisation practice. These have to be done at the same time every day and sent off to my supervisor for comment. It is causing a few problems as I'm not that organised. I am really feeling schizophrenic now as its encouraging me to explore 'the deep regions of my unconscious', although from the work that I have done on the Kabbalah, I have found it to be a most effective form of psychotherapy, but this in not necessarily compatible with the rest of my life at the moment. The first year of the course is supposed to be a magical therapy time to 'know thyself'. I have been warned that many things will start to change as a result of being part of a 'contacted school' (Report from the field no 5, 1 February 1992).

Being a member of an occult fraternity with a specific timetable and lesson structure added a regular and organized aspect to my fieldwork, particularly in opening myself up to 'otherworldly forces' (see Chapter 3). However, after a 'bad' ritual experience, which I describe in Chapter 7, I began to question how much I was willing to subject myself to esoteric experiences for the sake of the research:

This [ritual] has raised some very interesting questions for me. Primarily how much I am going to involve my subconscious in these meditations and rituals . . . I felt very disturbed for a couple of days after the ritual and awoke one night to find my bed surrounded by evil in the form of demons. It was then that I decided I needed to spend more time on psychic protection! (Report from the field no 14, 1 November 1992).

Inevitably some magical rituals evoke strong emotions among the participants. How does an anthropologist deal with this when there may only be support for true 'insiders'? The role of my academic supervisor – to temporarily 'hold the place' of more detachment from time to time – was especially important. On occasion it felt as though my supervisor was my only link with ordinary reality.

Participating in magical rituals is an intense experience: both in time and in personal relationships. Most magical seasonal rituals are performed at the same eight points in the solar cycle, and lunar rituals are performed on the full or dark moons. This meant that while I was engaged with more than one witchcraft group my fieldwork would reach a peak of activity at such times. This is demonstrated by one Spring Equinox, when I finished a ritual with one wiccan group at 6 a.m. in the morning, went to bed for a couple of hours and arrived at the location of the second wiccan ritual at 12 noon the same day. Working like this did present problems in the sense that sometimes it was easy to forget the specific details of how each group practised. On one occasion I got into trouble with a certain high priestess for not performing the Fivefold Kiss (a ritual salutation) in the way that she believed was correct.

Performing rituals and engaging in magical practice fosters intense personal relationships, and I developed a number of close relationships with magicians. In addition, it is relatively usual practice in some magical traditions for information to be passed on sexually, and at one point in the research, towards the end of my fieldwork, I found myself getting emotionally involved with a magician. Although it does not form a part of prescribed anthropological training, sexual relations with informants are considered by some anthropologists to transcend differences of self and other (P. Caplan 1993:22–3); but, in my case I considered such an involvement not only to be inappropriate but also a further complication within a complex fieldwork situation.

There were disadvantages to working in London, particularly when performing witchcraft rituals outside. The sheer number of people meant that the possibility of being disturbed was an ever-present reality. I experienced one night ritual in a wood in the middle of London that was rudely disrupted by a drunken man who would not leave. His presence affected everyone, and eventually the ritual had to be abandoned. On another occasion, this time a feminist witchcraft ritual (described in Chapter 5) when all participants were naked inside a sweatlodge in the garden of a deserted house, someone thought that they heard an intruder near our clothes and bags, which contained money and car keys, etc. We thought we could see the dark shape of a person outside from the flickering of the shadows of the fire on the inside of the sweatlodge. We all screamed, wailed and shouted as loudly as we could to scare the person off. The problems associated with trying to return home without my clothes and my car in the middle of the night ran through my mind at the time; and I was greatly relieved to emerge from the sweatlodge to see that, even if there had been someone outside, our personal belongings were intact.

The ethics of an anthropological study of contemporary magical groups have greatly concerned me. I have endeavoured to explain to the magicians that I have worked with that my research stemmed from a personal interest in magic, which started with my initial involvement with feminist witchcraft in 1984. Although I carefully explained that I was doing anthropological research, my main intention was to become accepted by the people I was working with, so I soon abandoned formal taped interviews, because these confirmed my outsider status (see Luhrmann 1989:17). Such interviews gave me data on magical ideology; which was useful in itself, but not the essence of what happens during the human interaction of ritual. Instead, I gradually came to rely on building up a network of contacts with whom I experienced rituals within various magical groups.

This was laboriously slow to begin with. Most magicians are friendly but cautious about allowing unknown outsiders into their rituals. Gradually I became accepted, and by the time I finished fieldwork I felt very much at home in the general magical social scene. It had become a way of life (albeit a very difficult one at times). I learned to memorize small details of rituals from watching and listening to what people were talking about and doing, the actual ritual itself and what it meant to other practitioners and to myself. Before and after rituals there were always discussions about magical workings and general magical issues and how these related to magicians' lives. I either wrote out my field notes when I arrived home, or more usually, in the event of a late evening ritual, the following morning.

As has already been noted, the magical subculture in London comprises a network of groups and people, many of whom know each other. I got to know a large number of individuals in diverse groups, and was very aware of the importance of 'holding my own counsel' and not passing on information and personal details about people I interviewed. I experienced a full range of responses to my presence as an anthropologist. On the whole, most people were interested, helpful and willing to give me information and include me in their rituals. However, one or two were suspicious of me, and one witch refused to give an interview on the grounds that I might be a 'plant from the *News of the World*'. Another, for her own reasons, spread rumours about my allegedly inadequate interview techniques. For some magicians the idea of anyone – anthropologist or group member – asking questions was threatening to their worldview and was treated with suspicion and, in some cases, outright hostility. I did have a difficult situation with one wiccan whom I had been meeting and corresponding with for a few months. We had met up, on average, once a fortnight, but also had long conversations on the phone. I showed him draft thesis chapters, and he commented on them. Out of the blue one evening he handed me a thick file of his writing on Paganism, which he said I could borrow to look through. I took this as a great compliment, and thanked him. A couple of days later I was reading articles in the file and came across one on ritual. This was very interesting, and I phoned him to ask his permission to quote a couple of paragraphs in an academic article I was writing. He agreed, and we left it that we would talk about it later. The next morning he came round to my house while I was out and delivered a letter accusing me of being manipulative and wanting to use his work without credit. We eventually resolved the misunderstanding; but the incident was a reminder to me that gathering anthropological data can be an extremely sensitive business.

This study, then, involves a dialogue between academic anthropological dis-courses and an internal examination of the process and philosophy of magical practice. As an anthropologist I have sought to examine the internal process of finding the magical self and of travelling between so-called 'rational' and 'irrational' worlds with the aim of explaining the magical world as I found it in London during the early years of the 1990s. These intense experiences opened up a part of myself that I had difficulty in combining with the academic discipline of anthropology. Barbara Tedlock writes that the public revelation of participatory details of field-work experience is still considered embarrassingly unprofessional by some ethnographers – mixing significant knowledge and personal experience is seen to discredit the endeavour. She points out that: 'The implication is that a subject's way of knowing is incompatible with the scientist's way of knowing and that the domain of objectivity is the sole property of the outsider' (1991:71). This raises a paradox highlighted by Paul Rabinow:

> As graduate students we are told that 'anthropology equals experience'; you are not an anthropologist until you have the experience of doing it. But when one returns from the field the opposite immediately applies: anthropology is not the experiences which made you an initiate, but only the objective data you have brought back (quoted in Tedlock 1991:72).

However, this position has shifted since the 1980s, and ethnographers have explored the process of producing ethnographies that represent the anthropologist as self interacting with other selves. Tedlock claims that this position reflects a climate of epistemological doubt; it also reflects the fact that more women, middle- and lower-class ethnographers, and third and fourth world scholars are engaged in fieldwork. They have been spurred on, she argues, by critical awareness and the radical democratization of knowledge. The effect is that ethnographic work is now attempting to bridge the gulf between Self and Other by revealing both parties as vulnerable experiencing subjects. She concludes that the change in ethnographic epistemology is not an apolitical solipsistic exercise in exploring the Other to find the Self, but an exercise in becoming bicultural (1991:80). It is this biculturalism – of translating the magical otherworld to the academic world – that is the aim of this present work.

Plan of the Book

The next chapter presents the otherworld as the central theme of this book. It begins with an exploration of otherworldly reality and what this means for magicians. This is followed by an anthropological and sociological theoretical perspective on the study of magic, and a critique of Western rationalist approaches. The stage is thus set for two ethnographic chapters that detail my participant observation: Chapter 3 focuses on my experiences of training to become a magician, and covers learning Kabbalah and magical philosophies of spiritual evolution, while Chapter 4 outlines ritual practice in three witchcraft covens and explores the centrality of women to the practice. The otherworld is a source of healing, and identification with otherworldly beings is seen to be part of a therapeutic process of finding the magical self. Magical identities are formed through an interaction with the otherworld rather than constructed from social discourses of the everyday world, and Chapter 5 examines magico-psychotherapies that are incorporated in rituals as forms of healing. This chapter concludes by examining magicians' use and abuse of power. Chapter 6 is a study of gender and sexuality, and as already noted, the practice of magic may be viewed as a romantic reaction to the Enlightenment emphasis on reason and its association with masculinity in its explicit glorification of 'unreason' and femininity. Certain gendered assumptions, such as the notions that female body is the locus of magical power and use of the magical will, are examined. Magical rituals are a cathartic space where relationships between male and female magicians are played out as psycho-drama; but since the introduction of feminist politics through feminist witchcraft, they have become ideological battlefields of meanings in the wider subculture. The question of morality is addressed in Chapter 7. The magical self, when internally 'healed' or 'balanced', is seen to be a 'power house' of energies of the cosmos, with tremendous potential for good or ill. Ultimately morality is internal – it is determined by the individual magician's sense of what is right or wrong. In high magic, morality is shaped by esoteric Christianity, whereas witchcraft has a more pragmatic approach, but is not worked out in such a systematic manner. The book concludes by drawing the different aspects of the study together and calls for further research on the otherworld.

Notes

1. This spelling of magick with a 'k' derives from the work of the magician Aleister Crowley; it stems from *kteis,* the Greek for 'genitals', and for Crowley denoted the emphasis on *Koth* 'the hollow one', meaning female genitalia. However, there is little etymological basis for this ascription. I am grateful to David Phelps for bringing this to my attention.
2. Kabbalah, sometimes spelt Cabala, Qabalah or Kabbala, is a form of Jewish mysticism based on a symbolic glyph called the 'Tree of Life'.
3. Practitioners tend to call themselves 'Pagans', while those studying them often use the term 'Neo-Pagan'.

4. This band, which was renamed Inkubus Sukkubus in 1995 for 'numerological reasons', attracts a large Pagan following and regularly performs at the London Marquee Club as well as touring the country. In recent years there has been a profusion of new Pagan bands such as: Praying for the Rain, Die Laughing, This Burning Effigy, The Mediaeval Baebes, and The Witches.
5. Margot Adler encountered the same difficulty in *Drawing Down the Moon,* her study of Paganism in America, and wrote that it was not possible to compile accurate statistics (1986:107).
6. The Pagan Federation is an umbrella organization representing the interests of many Pagans to counter what it sees as 'misconceptions, distortions and untruths' about Paganism. It was founded in 1970 as The Pagan Front to promote freedom of religious expression and an 'idealistic brand of Green spirituality' (Pengelly, Hall and Dowse 1997:4).
7. The history of the Isian religion, according to Steven, stems from the Templar knights of the Crusades and originated in the goddess queens of Egypt. The Priests of Isis started in 1960 as a 'white brotherhood' and spread underground; it was formed by witches 'too old to dance the circle round'. Conduct and beliefs follow 'a knightly code of honour, truth and love to fellow beings to fight for justice and uphold the sanctity of women'. The followers of Isis pray to her and feel her presence, but do not invoke her: 'she would be pissed off to come down to us'. Steven said that Isis makes people feel better. It is like 'someone has taken away or clarified a problem. You feel more able to cope with the next set of problems.' He said that the Isian religion is more like a form of Christianity: her followers relate to Isis individually with prayers, rather than invoking her as goddess.
8. Cunning men or women were those who were skilful or knowledgeable in curing illness in humans and animals, finding lost or stolen goods, fortune-telling, and identifying witches and dealing with witchcraft (Sharpe 1997: 66–7).
9. For an extensive overview of relevant sociological studies on NRMs see York 1995.
10. This is because 'By that year, Eastern teachers had opened ashrams and centers and books had been published representing the various strains of New Age concern, and a self-conscious social movement began to emerge' (Melton 1986).
11 Marcus and Fischer note that contemporary experimental texts concerned with 'personhood', defined as a convenient label representing culturally variable experiences of reality, may be divided into three groups: psychodynamic texts – based on a Freudian systematic interrelationship between conscious under-standings of social relationships and unconscious or 'deeply structured' dynamics (for example Obeyesekere 1981); realist texts, which allow the ethnographer

to remain in a position of unchallenged control of the ethnographic narrative, and which draw their initial frames of analysis from the public common-sense world (for example the dominant legacy of British ethnography created in the 1920s by Malinowski and Radcliffe-Brown); and modernist texts, which refer to the late nineteenth- and early twentieth-century literary movement formed in reaction to realism, and which arise from the reciprocity of perspectives between insider(s) and outsider(s) entailed in the ethnographic research situation (for example Briggs 1970; Favret-Saada 1980; Crapanzano 1980) (Marcus and Fischer 1986:48–73).

–2–

The Otherworld

For magicians the cosmos is alive. Magical philosophy represents a cosmology that is essentially holistic, based on the notion that the magician (as microcosm) is the locus of the macrocosm and that subtle energies and forces, which are inherent within all matter (animate and inanimate), are freely interchanged between those humans able to recognize and direct them. The essence of a magical training is to open up the magician's awareness to these forces so that they can be channelled, mediated and controlled. The cosmos is seen to be alive with forces and energies, some of which exist in a time and space distinct from, but also very closely connected to, everyday reality – the reality ordinarily perceived by the five senses of the human body.[1] This area is commonly termed the 'otherworld'[2] and can be perceived by the human psyche when in an alternative state of consciousness; it is said to co-exist alongside and in very close proximity to ordinary reality. Other-worldly forces are often personalized as deities or other creatures such as animals, combinations of animals and humans, or spirit beings such as fairies. Western magic is a form of 'occultism', a practice based on homo-analogical principles or correspondences that express a living and dynamic reality, a web of cosmic and divine analogies and homologies manifested through the operation of the active imagination (Faivre 1989). Modern Western magic is focused on directing the otherworld, and can therefore properly be classed as 'magic' in the sense of 'an attempt to exert power through actions which are believed to have a direct and automatic influence on man, nature and the divine' (Cavendish 1990:1). However, it also incorporates a religious aspect of respect and veneration, as opposed to worship,[3] of the forces or energies of the cosmos, which are personalized as various gods and goddesses. Western forms of magical practice are at once similar to and different from what has been called witchcraft and magic in other parts of the world: they are different from non-Western magic in their religious or spiritual aspect and similar in the sense that magic forms, as Evans-Pritchard demonstrated in his work on the Azande, the basis of a cognitive framework for knowledge. Thus Western magical practices are best described as magico-religious: they combine elements of both magic (control) and religion (veneration of deities).

The problem is that anthropological theories that stem from the post-Enlighten-ment rationalistic tradition cannot explain magic in its own terms, and thus do not pay sufficient attention to the otherworld. Anthropology was the first social science to point out that reality is culturally constructed and that there are multiple ways of experiencing the world; but anthropology is itself a child of Western culture, and it has given little credence to informants' accounts that do not accord with some rationalistic worldview of Western science (Young and Goulet 1994). Anthropology has a history of distrust of magic. The aim of this work is to place the otherworld and an understanding of magical philosophy in the forefront of the analysis.

Learning how to practise magic orders a magician's perceptual framework; it involves moving from experiential and intellectual incoherence or chaos to coherence and order in interpreting and 'mapping' the otherworld. This frequently involves learning to use mythology and metaphor as a language of the otherworld. The crucial requirement is that questions about this reality should be asked by a scientific discipline of anthropology that is prepared to extend its area of analysis to include quasi-universal human experiences that fall outside a rationalist framework. Studies of magic that do not engage with the otherworld dimension of contemporary practice cannot hope to explain magic, and it is inappropriate to use methods developed for the study of everyday reality to analyse the magical otherworld. Just as the current scientific method is largely based on a rationality grounded in a logic associated with linear causality, so the otherworld is governed by its own logic and must be studied in its proper context.

This chapter opens with an exploration of the otherworld. I then move on to look at the role of ritual in magicians' engagement with otherworldly reality. Ritual is a multi-modal symbolic form (Kapferer 1991), and my analysis is focused on its important role as a scaffolding for access to the otherworld. Finally, I outline the problematic relationship anthropology has had with magic.

Otherworldly Reality

Most human societies have some conception of otherworld/s[4] or a mode of reality alternative to everyday experience, which may be where the ancestors reside. Otherworldly reality is experienced through a shift in consciousness. It involves trance states and 'opening up' to a rich imaginative inner world that is often initially created during childhood. The clinical psychologist Richard Noll (1985) draws a comparison between shamanism and Western magic and asserts that both seek to control the spirit world by using the power of the imagination to obtain states of enhanced visual mental imagery; both shamans and Western magicians use mythologies as cognitive maps to structure their otherworldly experience. Magicians are usually very imaginative and creative people (for accounts of the role of fiction in creating imaginative otherworlds see Luhrmann 1989 and Harvey 1997). A good example is Jane, a feminist witch, who describes her childhood experiences during the early 1950s:

As a child I lived in two worlds – 'this side' and 'the other side'. This was conscious from the age of six or seven years old. I can remember being taken to a performance of the 'Nutcracker Suite' ballet. I was totally enchanted by the images, particularly the 'snowflakes'. I took these images home and lay in bed in the dark, watching them again and again. Eventually I became totally absorbed into them, and they took on a life of their own. I became a little human 'hoover', absorbing images from anywhere, including stories from comics and books, and took them over to the 'other side' where they became lived experience, and took on their own life. Similarly, if I had a particularly vivid and interesting dream, I learnt to go back into it, and it would take on a life of its own.

The otherworld becomes a place of total absorption, where otherworldly beings have a life of their own in an alternative reality that is closely linked with dreams. Jane describes how the otherworld – what she terms the 'other side' – also offers alternative ways of living:

I had warm and close relationships with people on the other side, often in unconventional, family-like groups, particularly being part of large groups of siblings, but with no parents around. I was able to experiment with life-styles. I was fascinated with the idea of royalty, of having servants, but in the end I decided that never doing anything for yourself made you stupid, vulnerable and helpless. So I came to prefer total independence, and in having friends and not servants.

The otherworld is a secret world with flexible boundaries:

On mornings when I wasn't going to school, I would go down to the bottom of my bed (perfect sensory deprivation conditions) and close my outer (this side) eyes, and open my inner eyes, and I would be there, on the 'other side', part of one of my lives. Sometimes the boundaries were crossed, people would come over and chat to me if I was on my own, and when I was out in the countryside, I would move in and out of woods 'this side' and 'the otherside'. I was always aware which side was which, and learnt to move with facility.

However, involvement with the otherworld may not be shared with anyone else, and Jane learnt not to discuss her experiences: 'I also learnt not to discuss this [the other side] with other people. I very occasionally tried to talk with a 'best friend', but soon realised that they did not understand, and so kept it to myself.' The secret otherworld can form the basis of a strong emotional attachment between magicians; it is one that is mediated by group dynamics, as is discussed in later chapters. Intense imaginative capacity is the basis of communication with the otherworld – moving from this side to the otherside: of becoming a specialist in the control of the process. It demands, as Jane notes, learning to 'move with facility' from this world to the otherworld, and utilizing such childhood experiences.

Magical training is about re-awakening this process, and it often concerns what is termed 'shape-shifting'. An example of shape-shifting, in a magickal magazine called *Talking Stick*, is given by Peter Wilkinson, and is titled 'Calling the Beast Within'. It gives the following instructions:

> Start to picture yourself creating a creature. I shall use a horse as an example. Try to create it around yourself, clothing your spirit in its flesh and blood. Remember that you are not invoking a separate entity but you are striving to be this creature. The sensations can feel peculiar at first, your perceptions may change: as a horse you have a wider field of vision and you are probably not seeing in colour any more. You probably will not be able to rationalise this at the time and it will come to feel 'normal' very quickly. As you continue, you might suddenly find, as many people do, that you are there. It is not really necessary to build up muscle layer on muscle layer in a perfect anatomical model. Often I have found that just by thinking and willing myself 'I am a horse', I am (Issue XI Summer 1993).

The development of the imagination is considered by magicians to be the most essential magical tool. It is the means whereby the magician thinks and feels herself into the otherworld and becomes one with the beings that inhabit its realms. The boundaries of self are broken and the magician becomes whichever otherworldly being she or he has embodied. Extraordinary beings are encountered and magical deeds performed in this alternative reality, and it may be a world of make-believe. One wiccan told me that the otherworld has a strong relation to a playworld, not of theatre play, but of games and joking relationships, and that it is often fuelled by fantasy. Fantasy re-enchants the world for many people, 'allowing them to talk of elves, goblins, dragons, talking-trees and magic'. It also 'encourages contemplation of different ways of relating to the world' and counters the 'rationality of modernity which denigrates the wisdoms of the body and subjectivity' (Harvey 1997:181–2).

Enchantment may exist side by side with the processes of rationalization. In an examination of computer games, Daniel Martin and Gary Fine argue that technology has become successful at creating fantasy, and that the current popularity of computer fantasy role-playing games, whereby participants venture into realms of mystery and enchantment, is evidence of a need for excitement and adventure (1991:121). Such games offer a mystical and subcultural opportunity for theatricality and identification; they also represent a decisive moral realm, featuring battles between good and evil. Problems posed by modern life – the recapturing of meaning through expressivity, creativity and imagination – are solved by venturing into a world where people can take temporary leave of everyday concerns within an adventurous alternative symbolic universe. The otherworld often functions in this way for magicians, many of whom enjoy such games, and who frequently describe their practice as a return to a sense of childlike wonder – a re-discovery of that which has been lost; but unlike fantasy computer games magical practice also involves live ritual enactment, which has a spiritual component. In magical practices the otherworld is a resource that, as the wiccan Ken Rees explained to me, is a space for physical, spiritual and psychological regeneration and nurturing; it cannot be replicated and is 'always open to amendment'.

Thus the otherworld is also the inner world; it is both internal and external – a combination of personal and social experience that involves a paradox of going out of the self to find the self within – and is specifically different for everyone.[5] The otherworld exists within the self, as Jane explains: 'The otherworld concerns the functioning of the human brain as in virtual reality but within me and with all my senses involved. Something very strong is happening within me, a very strong sensory experience which is the otherworld, although externally nothing is happening.' The otherworld involves a shift in consciousness. According to Carol, who has been involved in feminist witchcraft, the otherworld is:

> A land of everything out there and in here too [pointing to herself]. It is a whisper away, it is a shift in consciousness to see the bigger picture and the threads that weave through everything that has ever existed. It is multidimensional, a heartbeat away. There are people who exist in human form who have never lived in this world but have become ancestors. They live in their realm. They are who you meet on a [shamanic] journey. They are travellers too in a magical land, in parallel worlds. You can also shift consciousness to see the spirit in this room, the life force and auras, they speak to you through smell, pressure on your body, through vision, not only eye vision, but many different visions. The spirit residing in me merges the otherworld with this world.

Thus the otherworld is another dimension of this world, and otherworldly realms are also associated with spirit beings. It is viewed as another dimension both inside and outside the magician, and is stuctured according to different mythologies and magical traditions. In Chapter 3 I show how the otherworldly universe is framed by the glyph of the Kabbalistic Tree of Life and ideas about spiritual evolution, and in Chapter 4 I demonstrate how witchcraft generally has a more pragmatic approach shaped by notions of nature. Points of entry to the otherworld may be via sacred sites such as Glastonbury Tor (the sacred isle of Avalon), Avebury, Stonehenge, New Grange or the Rollright Stones. These sites are frequently seen as 'doorways' from the everyday mode of reality to the otherworld, and are connected with a mythological primal time which shapes the present (Bender 1998).

The otherworld is associated with magic; it *is* magic in the sense that in an alternative state of consciousness the magician experiences everything as connected and in relation to everything else. For magicians magic is energy, and practising magic involves moving energy. Leah, a feminist witch, explained to me: 'Magic is there, energy is there. Magic has to do with people finding exactly where energy is and shifting it, making it move. That's the magic. One just focuses it, sees where it needs to go and sends it there.' Magical energy is seen to be in everything that exists, and magical practice concerns channelling and using that energy. Magicians see the otherworld as a separate, although ultimately linked, sacred area; it is a place where it is possible to contact magical energies of the cosmos – the greater whole. The otherworld has its own reality, and otherworldy beings have their own existence and energies, which may be tapped into.

The common uniting link between all the diverse magical practices is the connection to the otherworld and otherworldly beings as a form of communication that can effect some form of transformatory experience. This might be quite informal. Leah, for example, told me that she was not 'out' publically as a witch, but that she worked magic as a storyteller. She told stories at Pagan and other events as spells to cause magical transformations in people:

> There are people in my coven who would answer [your questions] more definitely but I work differently; that's why I tell stories. If I choose the stories I can effect change without anybody knowing what I'm doing because stories are spells, they're changing on two levels, working on the conscious and the unconscious. I don't have to say to people 'I'm working on your unconscious now', I don't have to say that. If I've done my work right something will happen; and it does, it works. I will see the look or someone will go 'ah!'. My stories are almost all about transformation in some way, they are quite deep. That is the sort of work that I can do because I'm doing what I don't appear to be doing; and that's how I deal politically with the Craft. I do what I don't appear to be doing. I have a lot of Scorpio in me, secretive. I believe that myths work utterly.

Storytelling was Leah's way of effecting deep internal exploration as a spiritual transformation; another more public form of individual change is brought about by initiation. The magical initiation process may be summed up as follows: firstly, the individual has a personal experience – sometimes induced through a formal initiation, such as was practised in the classical Greek mystery tradition of Eleusis – which, it is held, cannot be told or spoken about, because it has to be experienced; secondly, this personal experience represents a shift in consciousness of 'finding the self.'[6] This involves a paradox: 'finding the self' leads to 'losing the self' (in the sense of the individual ego) in the greater whole. As one Pagan explained it to me, 'as I become more me, I become less me'. Losing the self enables the identification with and the embodiment of otherworldly beings as an expansion of self in 'shape shifting'; finally, this process leads to certain changes in the individual, which may not be dramatic but may entail slow, pervasive insights that effect a basic change in worldview. This is often interpreted as having spiritual significance.[7] A Pagan may recount having various transformatory experiences that may change how ordinary reality and identity are viewed.

Contemporary Pagan self-identities are organized around a deep internal exploration of the self through an interaction with the otherworld; the whole magical self is shaped through an intense emotional interaction with the otherworld (see Chapter 5). An example is a guided mediation at the 1996 Fellowship of Isis Celebration in central London, and represents a shift from postmodern fragmentation to initiation and rebirth. The speaker at the Fellowship of Isis celebration was the high magician David Goddard. He spoke about the Western Mysteries in terms of Egyptian mythology and the goddess Isis. We were all Osiris because we had all been 'dismembered, scattered, lost, hurt and abused'. Isis was our saviouress; as she had gathered up the fragments of Osiris after his murder by Set, so she would make us whole. Goddard described Isis as a 'manifestation of the eternal force of pure spirit', 'the womb in which all things live', 'the field in which we have elected to come to express our god natures'. The gods were described as a way to approach the 'supreme source which pervades all things'.

The whole conference of a couple of hundred people engaged in a guided visualization as we 'visited the Temple of Isis' in an altered state of consciousness. David Goddard used vivid visual imagery and rich description, including symbolic icons such as crystal ankhs (symbols of eternal life). We entered the world of Egyptian mythology by a camel, which had arrived to transport us to a statue of Anubis (the jackal god who is guardian of the Underworld). On arrival at the statue the camel disappeared and we were escorted inside the statue, which opened out into a colourful universe of beauty: 'the golden rim of the sun bursts over the eastern horizon and the sun's rays burst over us'. Individually we came face to face with Isis as our saviour and redeemer. We were left to reflect on the experience.

We had been invited to change consciousness to enter into a mythological Egyptian world – another reality, another time and another period – the ideal aim of which was to transform the self. This was an emotional and reflexive use of myth and pathworking to shape otherworldly experience: we were 'lost and scattered', but through the myth of Isis and Osiris and the mythological symbolism of the crystal ankh representing eternal life, and the sun bursting over the eastern horizon indicating initiation and rebirth, we were led to the source of the One – the saviour goddess Isis. The message is clear: fragmentation and meaninglessness is transmuted into wholeness, light and life.

Magical experiences such as these prompt questions about the ontological reality of the otherworld: whether it is purely psychological, or whether it should be grounded in a wider spiritual or religious framework. Magicians with whom I have discussed this are clear that there is a great difference between psychological and what they term spiritual explanations of magical experiences. They say that psychological explanations do not go far enough, because they can only be attributed to an individual. For them there is a greater whole – a 'something' organizing the experience. I asked what the difference was, because, using the microcosm/macrocosm analogy (that the microcosm is a part of the macrocosm and vice versa), the psychological was a part of the whole anyway. One magician responded that it came down to our culture and how different scientific disciplines viewed the world in a partial and fragmentary fashion – the focus on the part precluded being able to see the larger dimension. The consensus for these magicians was that there was a wider energy field that was experienced in different ways according to individual psychology and cultural shaping.

A Source of Knowledge

One thing is clear: the otherworld is a source of knowledge. Edmund Leach saw the essence of ritual located in a relation between the world of physical experience and an otherworld of the mystical imagination that acted as a channel of communi- cation for making the power of the gods available to human beings (1976:82).The central defining feature of a magican is being one who communicates with the otherworld, to engage with the beings therein and mediate the forces bringing back knowledge and wisdom to ordinary reality.

In high magic angels, as otherworldly beings, are regarded as a source of divine knowledge. Viewed as intermediary forms of consciousness between the magician and other forms, they are seen as messengers or mental, astral or etheric waves or radiations. Angels or archangels are convenient labels for 'superhuman beings whose "aura" or wavelength can be contacted by the subjective methods of inten- tion, concentration and desire'. The high magician David Goddard describes an angel as a 'being of pure consciousness, unlimited by time or space' that 'bathes in the radiant energy emanating from the Godhead'. Each angel is a focus of power, wisdom, beauty and love. An angel 'channels the divine light without distortion, and functions in utter accordance with the will of The One. It has no will save that of God' (1996:xvi). For Goddard, angels function to lay the lines of force: the all- pervading energy of The One is focused by the angels into patterns. One such pattern is the Qabalah (Kabbalah), which was given to humanity by the angels. Angels also have a teaching function: Michael assists in matters of achievement and ambition; Gabriel is concerned with the development of psychic abilities; Samael, a protective angel, is invoked for courage or perseverance; Raphael's speciality is healing and communication; Sachiel's rulership involves financial matters; Asariel rules all forms of mediumship and trance work; Haniel is concerned with the devas of nature and the 'Hosts of Faerie'; Cassiel supervises the energies of the planet; and finally, Uriel is the transmitter of the Magical Force itself (Goddard 1996:2–17). Angels are thus intermediaries between the magician and divinity – they are spiritual messengers, and can be invoked as sources of wisdom.

Initiation is believed by magicians to mark an especially important learning experience – as an opening up to the wisdom of the otherworld. In witchcraft this is often a two-stage process. Initiation rituals, as *rites de passage,* expand conscious- ness to the non-ordinary realities. Healing typically involves the destruction of the old identity or ego, and is often experienced as a crisis. This is often related to, and experienced as, the Sumerian myth of Inanna who journeyed into the otherworld (in this myth it is termed the Underworld) to find Ereshkigal, her dark sister, who is really the repressed part of herself (see Chapter 6). The goddess Inanna is what Perera (1981) calls a many-faceted symbolic image representing wholeness. She combines earth and sky, matter and spirit. This myth introduces important magical ideas: it encourages an exploration of the self; it introduces the notion of a dialectics of change – nothing is static in the movement from ordinary world to otherworld and its connection with the changing seasons of the year; and it encompasses the notion of the unconscious as a realm, a universe to be charted, explored and finally understood. This last aspect concerns the initiatory process, which essentially means discovering the unity with nature and the cosmos and the rhythm of life, death and rebirth. The descent to the Underworld is when the witch has to face who she really is – she has to face her own anger, fear, destruction and death.

Opening up to otherworldly reality is a process that requires learning to interpret one's being within the wider magical whole – learning to see connections between the planetary forces and the self. This leads to a negotiation of identity – the formation of the magical self – the identification of the self in relation to the cosmos (see Chapter 5); it also involves learning to communicate with otherworldly beings that are often anthropomorphic – like gods and goddesses – or zoomorphic – like animal spirits or 'familiars'.

The second stage of initiation frequently involves working with otherworldly beings, learning how to use their wisdom, and to channel their energies within the microcosm of the body. Experience is gained through ritual and the calling in of the spirits of the four quarters of the witchcraft circle: east, south, west and north. These correspond to intellect, will, emotions and the body in the human microcosm. In addition, the witchcraft solar cycle of seasonal rituals, the 'Wheel of the Year', present an opportunity to invoke different gods and goddesses who are appropriate to their season – for example, Brigid, the Goddess of fire, inspiration, healing and poetry, is associated with Imbolc which is on or near the second day of February, and Lugh, the Celtic sun god, is celebrated on the first day of August, as his power begins to decline before he is reborn at the Winter Solstice.

In many Pagan practices ancestors are seen as a rich resource of knowledge. The druid Philip Carr-Gomm argues that when each generation stands on its own and 'doesn't feel connected to its lineage' it experiences alienation and disconnection. But, conversely, when it is in touch with its ancestors, 'when we sense them as living and supporting us, then we feel connected to the genetic life-stream, and we draw strength and nourishment from this' (1991:82). In witchcraft images of older powerful women as crones or hags are very important, because they are seen as a connection to the ancestors. The Goddess is seen to be a triple goddess, and she passes mythologically from maiden to mother and then on to crone before starting the cycle again. In this mythos hags and crones have great power, and are seen as sources of wisdom. They are often associated with insight, the earth, dark, blood, death and also birth, because death is seen to lead to life. Jane, the feminist witch, described an otherworldly experience of meeting an ancestor in the form of a Kalahari San woman at a weekend workshop run by Starhawk in 1989. The trance journey took place in a ritual space where a large group of women had been divided into several smaller groups, and she describes the experience as follows:

> I was in a group of five women. Sitting, holding hands, we shared a group trance where we were in a boat at sea. It was a small boat, and we were laughing and joking together. We finally came to shore, where we all got out and pulled the boat out of the water, and then walked off along the shoreline, separately and in silence. We had been asked to seek out the Crone, the Hag, and gain some information from Her.
>
> My beach was pebbles, shingle, and sloped up to fairly low cliffs of yellow-brown sandy material. It was about late afternoon, as the sunlight was slanting at an angle. There was a cave entrance in the cliffs, and I went in and walked through a long tunnel which was occasionally lit up by sunlight coming through cracks. Eventually, the tunnel opened into a larger space. I felt a ledge with my hands, and sat down in the dark and waited.
>
> My eyes got used to the dark, and I could feel a presence in the opposite side of the space to me. I could hear a soft noise, and then some light came through into the dark, and I saw her, squatting in a corner, laughing very quietly, a San woman, very old, her face full of wrinkles like someone who has lived their life out of doors in a hot dry climate, wearing a leather 'apron' and leather *kaross* [cloak] with ostrich egg beads around her neck. She was squatting in the corner laughing at me – not unkindly, just amusement. I sat there, feeling surprised, as I had expected a 'classic' Crone, European, long white hair, frightening, awe-inspiring, stern, and probably telling me off, telling me what to do – not shared amusement. We sat there together, the ancient African woman and I, and then I was called back to the ritual space.

Jane said that this encounter with the African woman taught her that 'life was an amusement and that nothing was serious'. This was a survival technique for getting through old age, which was difficult and painful at times. She said that everyone had traumatic experiences, but the insight provided by this black African woman as ancestress was an 'enormously appropriate' way of dealing with getting older – 'in the end it passes, and sitting and laughing puts everything in perspective'. Thus the otherworld can provide a rich source of wisdom that helps individuals deal with the everyday realities of this world.

Ritual and Otherworldly Reality

Individuals' experiences and expectations of the otherworld are often arbitrary and unstructured until they are shaped by the process of engaging with a specific magical tradition and training (see Chapters 3 and 4). Ritual is an important part of most magical traditions, because it is seen to be a holistic healing space from the everyday world where the magician can contact her or his inner world and the wider forces and energies of the cosmos. Alan Richardson expresses the essence of magical ritual from a high magician's viewpoint:

> We are so used to the modern age with its projected two-dimensional entertainments, electric lighting, double-glazed and centrally-heated lives that we have forgotten the impact of the great rites. We have forgotten the effect of candle-light and shadow, chanting, incense, ritual and invocation, when each member is robed and masked, unknown, a pair of glittering eyes expressing power in a throbbing room. Used to the cult of the personality we have forgotten the awe that such obliteration of the face can bring, forgotten the way that we ourselves, when cloaked in anonymity, can become something other than our normal selves, momentarily greater and truer to our higher natures, no longer tied down by physique or physiognomy but pure expressions of energy. In such rites the magician feels himself and his world charged with immanence, with the sense that some great door is about to open between the worlds, and the gods pour through; he feels as though his mind is peeling open like a flower, revealing the jewel glowing within, glowing with the *spiritus mundi* that dwells inside, awaiting its call (1987:113).

Ritual creates the space where what is seen as the ordinary self gives way to something greater. Contact with a greater power is achieved through an exchange of energy. Ritual harnesses the magician's latent imagination and creativity produced through an alternative mode of consciousness; it directs and focuses it, through the will, to a specific objective.

Recent work in anthropology has attempted to explain the biopsychological processes that produce the other mode of consciousness associated with an experience of the otherworld. Charles Laughlin argues that ritual techniques generate and distribute psychic energy and that there is a cross-cultural invariance in psychic energy. Building on the work of Stanislav Grof (1976) and Kenneth Ring (1976), Laughlin used his seven years experience as a Buddhist monk to argue that, despite different cultural and symbolic traditions, there is a 'recognizable structural invariance' in the reports of higher psychic energy experiences. In Tibetan Tantric Buddhism, as in other esoteric practices, the body is seen to be made up of a system of channels through which psychic energy passes and ritual techniques establish a free flow of energy. This is the pre-requisite for the expansion of consciousness. Laughlin argues that patterns of sensorial psychic energy may be either the consequence of metabolic events in the body or they may produce metabolic events in the body, so that psychic energy is felt as bodily sensation or 'seen' as visions of energy flows. He maintains that the apparent diversity of psychic energy experiences is due to cross-cultural differences in symbolism and interpretation; but the underlying function of the human organism is the same (1994:129). Ritual techniques and experiences open up the magician to the otherworld.

Classical anthropological studies of ritual, while not being concerned with such experiences, are nevertheless valuable in understanding ritual. Durkheims's demarcation of the sacred/profane dichotomy influenced Arnold Van Gennep's tripartite schema of separation, liminality and re-incorporation (1908). Durkheim's dichotomy is useful to analyse the special ritual space that magicians carve out from ordinary, everyday time in order to make contact with otherworldly forces. Edmund Leach has pointed out how ritual serves to express a relation between the world of physical experience and an otherworld of the mystical imagination. The ordinary world is inhabited by mortal humans, who are often powerless. By contrast, the otherworld is inhabited by immortal gods, who are powerful. Ritual then provides the channel of communication by which the power of the gods is made available to humans (1976:82). The fundamental prerequisite of any magical ritual is the creation of a special space – a space and time 'between the worlds' where the magician can connect to the otherworld. Durkheim's sacred/profane dichotomy contrast as developed by Van Gennep, Leach and others is a useful analytical tool to understand the way that magicians create a specific space to gain access to the otherworld where communication with otherworldly forces, personalized as gods and goddesses, is possible.

Many anthropologists have understood ritual symbols primarily by their social function: for example, Mary Douglas (1966). Douglas viewed the physical body as a microcosm of society, and the body was also the locus of moral values exhibited in witchcraft beliefs (1970). Geertz, on the other hand, was more concerned with searching for the meaning of symbols, but the construction and utilization of symbolic forms lay out-side the individual in an intersubjective world of common understandings (which induced psychological dispositions) (1973). Both saw humans imposing experience on systems of classification that were ultimately built out of social materials. Other anthropologists have attempted to bring the psychological (the internal) into an analysis of ritual symbols; but ultimately this too is explained by the social (Victor Turner 1967). Perhaps Gananath Obeyesekere has gone the furthest in trying to understand the internal workings of the human psyche. Like Geertz, Obeyesekere draws on the interpretative sociology of Weber, who saw culture as the result of the human tendency to impose meaning on existence. However, he was dissatisfied with the way that the interpretative tradition has ignored the unconscious. Obeyesekere (1968) demonstrates how cultural symbols are articulated with individual experience – personal symbols and cultural symbols operate individually and at a cultural level at the same time. The complex psychological experiences of the individual coalesce around pre-existing meanings that are imposed by culture.[8] Obeyesekere's analysis incorporates the unconscious; but the baseline is still the social.

Working to extend understanding in the area of ritual, Edith Turner draws on her late husband Victor Turner's writing on Ndembu ritual symbolism. Edith Turner points to the way that he goes beyond the idea of symbols as abstract referents to see them as a unitary power in which is conflated all the powers inherent in activities, objects, relationships, and the ideas they represent (Victor Turner 1967: 298). In her own re-examination of the Ihamba curative ritual (where the doctor draws out a harmful tooth from the back of the patient) she goes further in her analysis by drawing on the anthropologist Michael Harner's work on shamanism (1990[1980]) to explain the awareness of more than one level of reality operating in the ritual. She argues that it was the unitary power of *Wubinda* [the Ndembu spirit and cult of huntsmanship], combined with medicines, the participation of five doctors and a crowd, and the doctor's tutelary spirits that drove into visibility the spirit substance that she saw during the ritual as a 'gray blob'. Edith Turner is here describing a different form of reality – what may be considered to be an alternative form of consciousness – a different cognitive framework for knowledge (1992:92).

In the Western context, rituals function as a scaffolding for contact with the otherworld; ideologically they offer the individual practitioner a direct route to self-transformation through an experience with the otherworld. Contemporary magical practice is diverse, and there are important differences between witchcraft and high magic (see Chapters 3 and 4); but there are certain basic features of ritual that remain essential to both, and indeed all contemporary magical practices. These are: firstly, the creation of magical language in ritual as a symbolic system and as a means of bringing through magical power from the otherworld; secondly, the emphasis on the body; and finally, the development of the magical will. I will briefly deal with each of these aspects in turn:

Creation of Special Language

In contemporary magic, ritual is the space for contacting the otherworld. The use of special language clearly demarcates the ordinary from the otherworld; and it also indicates otherworldly powers for magicians. Much magical practice is conducted in a ritual circle. This practice stems, as Norman Cohn has noted, from the historical ritual magic practice of conjuration of demons and the coercion of spirits (1993[1975]:100–4).[9]

Much modern magic is thus based on the creation of a sacred space from the mundane world, and there is a good deal of technical language that is related to this central aspect of the practice. Magicians talk about 'casting' or 'opening' a circle, of 'calling in' the elements, the forces, or the light. What they are doing is demarcating the ordinary world from the magical otherworld; and in this process they are changing consciousness and preparing to enter a different world. When a magician is in trance he or she usually feels 'at one with' otherworldly forces: there is no distinction between the individual magician and the cosmos, and the magician 'embodies' cosmic energies. When the ritual is finished the reverse procedure is adopted by the 'closing' of the circle and the return to the ordinary world, which is seen to be accompanied by 'grounding', the re-establishment of normal everyday consciousness.

The ritual circle is also an 'amplification chamber'[10] for the powers of the cosmos, which are channelled by magicians' invocations and spells. Malinowski noted that words used in spell-making had a mystical power associated with mythology, the influence of the ancestors, and a sympathetic influence of animals, plants and natural forces.[11] The ethnographic work of both Jeanne Favret-Saada (1980) and Paul Stoller (1987) also emphasizes the importance of magic as a language system for explaining and manipulating such mystical power.[12] The role of esoteric language is thus important for magicians to create a special space for the communication and channelling of otherworldly energies.

The Emphasis on the Body

The body is the initial focal point of all Western magical work. In contrast to 'exoteric' Christianity (which has viewed the body as sinful and in need of control) the body in magical practices is sacred and the site of the macrocosm. This means that the individual body contains all the energies and forces of the cosmos. It is a field of energy that has the potential to channel the forces of the entirety of the macrocosm. The body, for magicians, is understood through an ongoing relationship with the otherworld through magical practices. For magicians the body is the route to transcending ordinary consciousness (although there are differences in approach between high magic and wicca). Working ritual involves fully experiencing the otherworld through a shift to an internal frame of reference.

This process can be illustrated by a magical technique taught to me at a 'Circlework' workshop for beginners learning magical practices. We were taught a 'North American Indian technique' of greeting and introducing ourselves to the quarters of the circle. This consisted of sitting in the centre of the circle and addressing ourselves to the four quarters, one by one. The remaining members of the group spread around the quarters, moving from time to time to get the different feeling that each evoked. The person/people in each quarter were to ask questions that were appropriate for that quarter of the person in the centre. A magical circle in Britain is always based on the following 'correspondences': the east is concerned with rational thought; the south with will and energy; the west with emotions; and the north with the body. On this occasion each person took a turn in the centre of the circle and had a chance to think and experience what that quarter represented to them. One man described his feelings about his body – that he ate and smoked too much at times. Someone in the north spoke back to him as his body – 'It makes me feel bad when you make me smoke too much, can you cut down a little?' At the same time, the north, south, east and west each also represent different aspects of the human body; all is experienced through the body. Even rational thought (the east) is experienced in relation to the body, the will and the emotions: therefore it can never, in theory, become disembodied and split off as spirit from inferior matter.

Development of the Magical Will

The magical will has been a central defining tenet of magical practice since the Renaissance (Yates 1991). It is commonly understood by magicians to mean the direction of the mind and the emotions to a particular magical objective. The magical will represents the magician's power, and is seen to 'vitalize' the imagination (Regardie 1991b:126). In short, the magician has to balance the powers and energies of the macrocosm within his or her body (the microcosm) and then directly channel them, through the active imagination, to a particular magical intent. The magician is the form into which the energy or force is channelled. This is clearly demonstrated by the high magic Abra-Melin ritual, where the magician uses the force of evil within to energize the good, which ideally then transcends the evil. This ritual takes six months to perform, and the climax is reached when the magician meets her/his Holy Guardian Angel, who is the 'higher self' and who, according to one magician, will have 'bestowed grace and splendour upon his soul, sustenance into his spirit, and flooded the whole sphere of the mind with an all-encompassing illumination, which no words may adequately describe' (1991b: 195). After this the high magician proceeds in the evocation of good and evil spirits (these are often described as personal neuroses or 'complexes' – 'subconscious powers' or spirits) with the intention of 'subjugating' them, and 'consequently with them the whole of Nature, to the domination of his transcendental Will'. The high magician believes that he or she uses the sacred forces of good and evil to purify him- or herself. In this context, the profane is analogous to the ordinary everyday world. The ideal aim of ritual practice is personal transformation, and this often interpreted as a spiritual awakening that gives meaning and direction to a magician's life. It is also the source of magical identity, and I shall examine this in Chapter 5. From these features it can be seen that ritual provides a special space for communication with the otherworld; and in the chapters that follow I shall examine magical ideas and practices that focus on the themes of identity, gender and morality.

Anthropology and Magic

The early anthropologist Edward Tylor saw the magical arts in civilized European societies as survivals from a barbarous past (1871). A similar view was taken by Sir James Frazer, who maintained that magic was the first stage in the evolution of the human mind, and that both magic and religion (which grew from the mistakes of magical thinking) would eventually be superseded by science (1993[1921]). Frazer worked on Tylor's view that magic was based on the association of ideas, and developed the notion of sympathetic magic – the belief that everything is connected in a state of sympathy or antipathy. Tylor and Frazer were interested in the evolution of ideas and not in magic or magical practices *per se*; nevertheless, paradoxically, Frazer's definition of sympathetic magic can still be taken as a basic definition today. Frazer argued that: 'Things act on each other at a distance through a secret sympathy, the impulse being transmitted from one to the other by means of what we may conceive as a kind of invisible ether' . . . (1993[1921]:12). This is the underlying theme behind communicating with the otherworld in the diversity of contemporary magical practices.

A sociological view of magic was taken by Marcel Mauss, who defined it by its social context and focused on the symbolic meaning of magical acts. Like Tylor and Frazer, Mauss, drawing on Frazer's notion of sympathetic magic, argued that sympathy was the route along which magical forces pass, but it was ritual that provided the power. Mauss marked magic off from religion: in a religious rite the individual was subordinated to forces outside his or her power (society), while in magical rites the individual appropriated the collective forces of society for his or her own ends (1972[1950]). The approach of Bronislaw Malinowski also took a sociological view of magic by utilizing Frazer's principle of sympathy – the knowledge of how to use words in magical spells had a mystical power in Trobriand magic, and ordered the social system. But for Malinowksi magic was collective, yet also had a psychological dimension because it resided in humans. It was not an abstract force, but was born of practical situations and emotional tensions coming into play when technical knowledge had reached its limits (1954[1948], 1978 [1935]). Mauss and Malinowski thus departed from Tylor and Frazer, who explained magic by intellectual beliefs and ideas; they took a more sympathetic approach to magic, but were ultimately reductionist in their views and do not further an understanding of the magical otherworld as such.

Social scientific studies of magic have been influenced by Durkheim's distinction between profane magic and sacred religion, which drew on Mauss's distinction. The sacred, for Durkheim, was represented by two religious forces: the pure, which was the beneficent powers of deities; and the impure, everything which was evil or disorderly such as death. The sacred was a force in society, and was 'created' by society. For Durkheim, religion originated from society. It created evil powers and at the same time provided expiatory rituals to diffuse and control them (1965[1915]). The value of Durkheim's approach, even though he took a negative view of magic, lies in his overcoming of the monotheistic good/evil dualism, and in his broad understanding of the sacred. The sacred may be likened to the holistic philosophy of magic – the idea of the individual's being totally integrated with the cosmos – achieved through the magical ritual. The sacred, for Durkheim, was a force, both pure and impure, that could explain evil and suffering – malevolent powers were collective *représentations* of suffering.

Magicians also use the term 'the force' for describing their connection with goddesses and gods, spirits and demons. This force can be used for good or ill, depending on the intention or will of the individual magician. Magical cosmology is holistic: therefore good and evil are both contained within it, and the latter is not excluded, as it is in Christianity, for example. The magician must understand both the good and evil energies within himself. However, contemporary magical practices may be more usefully defined as magico-religious movements that incorporate an alternative mode of thought. This last is associated with a magical otherworld, which in turn organizes social action/relations.

This work offers a challenge to the way in which anthropologists have examined magical practices, as a result of which the classical distinction between magic and religion falls away.[13] This book is a study of magical experiences gained from communication with the otherworld that are not recognized by rationalistic scientific methods. Tylor claimed that magic was a pseudo-science – false associations were taken for causality – and magic was a delusion based on the association of ideas. However, Frazer, Mauss and Malinowski all claimed that the logic behind magical thinking was rational. Evans-Pritchard's classic ethnography of the Azande, which explained witchcraft as a natural philosophy providing a coherent way of organizing Zande life, also called it rational, but states that in the final instance it failed to match up to European standards of rationality (1985[1937]). Thus European notions of rationality and the discourse of an often positivistic science have been used as a universal benchmark against which other cultures are judged. Richard Shweder (1991) has argued that since Nietzsche announced that God was dead all anthropological theory is atheistic by assumption, and that much debate on gods and witches starts by 'null-reference' reasoning – they do not exist. Avoiding direct analysis of 'magic', anthropologists have sought to discover whether the performance of a magical act is instrumental (whether the logical premises behind it are similar or dissimilar to those of a scientific experiment) or, following a symbolist approach, have examined whether a magician is trying to achieve the same results as a scientist, through using a different language or mode of thought (Tambiah 1979).

Evans-Pritchard's debate with Lévy-Bruhl over so-called primitive peoples' prelogical and mystical mentality, which was based on the 'law of participation', has been formative to studies of rationality in the discipline and beyond. Evans-Pritchard sought to prove that Azande magical beliefs were logical, rational and similar to Western conceptual frameworks if they were examined in their social context. His work became central in what was termed the 'rationality debate', in which philosophers analysed how far any belief was dependent on social context. The positions of relativism and universalism formed two broad schools of rationality. According to Stanley Tambiah, the 'relativizers', such as Winch, Geertz and Wittgenstein, claimed that there are multiple 'rationalities' or styles of reasoning and that translation of cultures is difficult but possible, and involves careful 'mapping' of other cultures' understanding on to our understanding, which may be modified in the process. By contrast, the 'unifiers', such as MacIntyre, Gellner and Lukes, argue that there is only a single rationality based on universally valid rules of logic and inference, and that modern Western analytical rules and concepts form the basis of categories of understanding. Comparative and transcultural judgements can only be made regarding the degree of rationality or irrationality manifest in a particular belief or action system (Tambiah 1991:116).

Magic as Pseudo-Science

Tanya Luhrmann, in her analysis of Western magicians' notions of magical beliefs, notes that a primary difficulty with the so-called rationality debate lies in its assumption that beliefs are clear-cut and coherent. She finds that this position has left the conception of 'irrational action' ill-formed and problematic. Consequently, she notes, there is a difficulty in arguing about belief arising from the way that the debate has centred around two opposed positions. The first is an intellectualist viewpoint, stemming from Tylor and Frazer, and is based on the notion that magic is similar to science – because it arises out of natural thought processes; however, it is erroneous, being grounded on false premises. The second is a symbolist position, attributable to Durkheim, which explains magic away as a 'romantic rebellion', neither rational or irrational. In explicating this latter view, Luhrmann draws on Skorupski (1976:18), who interprets the Durkheimian position as one of science's being distinguished from magic and religion in the same way as literal meaning is distinguished from symbolic meaning. Thus the two poles of the rationality argument centre around the intellectualist notion of magic as a pseudo-science and a symbolist view of it as an emotive and expressive practice that does not involve belief. Luhrmann argues that modern magicians' beliefs are not fixed or consistent, and that they drift in a complex 'interdependency of concept and experience' (1989:353). She claims that the debate between intellectualists and symbolists has cleared a path for a new sort of discussion and new ethnographies that describe the cognitive impact of cultural experience and that take the 'multifarious mind' as a topic for analysis. She asserts that the real questions behind anthropological work on ritual are old philosophical questions on the nature of mind and what is knowledge.

There are a number of difficulties with Luhrmann's contentions. The first concerns the issue of belief, and is based on her assertion that magicians' beliefs are not fixed. This shows a misunderstanding of the basic principle of magic, which, as I have argued here, is concerned with communication with the otherworld and the bringing through of power. Magicians clearly differentiate between this world and the otherworld, and ritual is the major means by which they contact other-worldly forces. This position was made clear by the wiccan Ken Rees when he explained it to me like this: 'For me Magical Space is primarily other and intentional. This space is deliberately constructed, prepared, consecrated and reinforced to be a reservoir for an amplification of power and resources (emotional, physical, spiritual). Once achieved with the skilled spellmaster, this space acts as a superior arena to non-magical space for the achieving of results.'[14] Rees points out that the magician should feel more empowered[15] at the end of a magical working, and that the action in magical space is intensified, faster and more focused than any equivalent intent in non-magical space. He claims that magic 'calls upon and utilises the same (re)sources as our everyday selves'; but in the magical space these are heightened. Rees makes clear the distinctions between this world and the other-world, profane and sacred, and notes that these are essential to the fundamental workings of magic:

> It seems to me hard to do magic without the operation and application of distinctions. Whether these are dualistic or more, are couched in the vocabulary of the sacred versus the profane, natural versus supernatural etc. is perhaps less important than the sense of difference that should arise upon doing magic, upon turning one's face to deity and addressing it/them in whatever ways are chosen. This clearly does not mean that magic cannot occur outside demarcated sacred space. We know it can and does all the time – but I do not think this is the same kind of magic. That magic occurring outside formally arranged settings is much more likely to be of a synchronicity nature, or serendipity, or may be the lightning flash. Intentionality does not come from the operator – indeed it might be hard to talk about intentionality at all in such contexts, rather random effects . . .

It is clear from this that the practice of magic concerns primarily the conscious intentionality of the magician in the creation of a special magical space to contact the forces of the cosmos in order to channel them and bring through power from the otherworld. Rees argues that magic can occur outside specially created space, but that in that case it is unfocused and undirected. Thus the intentionality – or the magical will – of the magician is crucial to the performance of magic. In fact, magic cannot be performed properly without the conscious focusing of the magi-cian's will. There is nothing inconsistent in this belief. This is the basis of magical working, but it does nothing to contradict the fact that magicians, as human beings, will also have contradictory, confusing and unfixed beliefs *as well* – this is the human condition. Luhrmann has not taken the magicians' ontological reality of the otherworld seriously, because it does not fit into her explanatory framework.

The second difficulty with Luhrmann's analysis concerns the nature of mind and what is termed knowledge. Her position is heavily dependent on a positivistic view of science: rationality must conform to knowledge based on observable facts, and this form of science is the only form of valid knowledge. In other words, otherworldly experiences – magic – do not form a basis for a scientific knowledge, for an understanding of the world. In this view magic must be irrational and necessarily false, because it does not conform to Western scientific criteria; she equates reason with a positivistic view of science. Luhrmann cannot examine the magical otherworld, because it does not fit into her theoretical framework. She therefore conflates magicians' experiences of the otherworld with the intellectualist tradition of magic as pseudo-science.

The framework for understanding what Luhrmann has called the old philosophical questions of the nature of mind and questions of what constitutes knowledge must be radically re-envisioned if we are to understand the practice of magic today. My approach is reflexive: I include my experiences of the otherworld within the fieldwork context as part of the process of understanding and analysing otherworldy knowledge. This type of methodology asks new questions of anthropology as a social science: it breaks down the conventional barriers set up between the anthropologist and the 'other', and it directly challenges the traditional approaches to anthropological knowledge that are framed by Western notions of rationality.

Beyond a Western Rationalist View of Magic

When it comes to the study of magic, a positivist paradigm has been common in the social sciences generally. Anthropologists working on magic and related subjects such as spirit possession and exorcism have searched for alternative explanatory and theoretical frameworks. Bruce Kapferer, in a study of Sinhalese exorcism rituals (1991[1983]), takes a phenomenological orientation that gives full weight to the understanding of human experience and does not give preference to rationalist social science theory as an exclusive and superior model for understanding the complexities of human life. Phenomenology sees all human orientations to reality as constructions, with no particular construction necessarily more 'true' or 'real' than another: 'all constructions reveal dimensions of the nature of human being and existence' (1991:xviii). A phenomenological approach is non-reductive: it does not reduce human experiences of the otherworld to social or psychological explanations, but instead involves a radical suspension of the analytical categories of normal rationalistic social science: 'It demands a turning back to

the world of lived human experience and taking what people do and say seriously'
(1991:xix).

Relativists, like phenomenologists, have always taken other cultures seriously,
to the point of over-emphasizing difference. Relativists have always argued against
the position that there is one unique truth, especially the truth revealed by a
positivistic science. They assert the reality of a multiplicity of knowledges, and
have been opposed to a hierarchy of knowledge in which a rationalistic view of
science is viewed as superior. In an effort to understand the shamanism of the
Piaroa of the Orinoco basin of Venezuela, Joanna Overing has drawn on the radical
relativism of the philosopher of science Nelson Goodman to explain the ways that
the *ruwang* – the religious and political leader of the Piaroa – constructs otherworlds
through time. Goodman's position, outlined in *Languages of Art* (1968) and *Ways
of Worldmaking* (1978), is to accept the reality of a multiplicity of knowledges or
versions of the world. His interest in structure led him to cross the boundaries
between art and science and to look at the creative processes followed in both in
their respective construction of worlds; there is a good deal of overlap between
science and art (in the use of metaphor, for example), and Goodman recognizes a
plurality of knowledges equal to science. In an earlier work on reason and morality,
Overing argued that the construction of metaphor is a natural extension of ordinary
thought, allowing the cognitive passage from familiar to unfamiliar. Metaphor is
a mechanism for changing the mode of representing the world in thought and
language, 'Through metaphor and the linking of incongruous concepts, one can
see things in startling new ways' (1985:5). She now calls for new means of insight
into the processes of the shaping of knowledges: '. . . new means, the more technical
the better, of acquiring insight into the processes of creating knowledges, both
generally and in their multiplicity' (1990:604).

Part of the problem of defining worldviews is inherent in the definition of science
itself, and current debates in anthropology, as in the other social sciences, are
being articulated within the currently fashionable context of postmodernism. Brian
Morris argues that postmodernism, which attempts to deconstruct science by
problematizing both instrumental reason and the association of truth with science,
is an exaggerated reaction against positivism (1997:315). He takes a stance against
the postmodern nihilistic deconstruction of science – the 'end of science' – and
calls for approach that combines science and hermeneutics, but which avoids
the extremes of positivism, by pointing to scholars such as Ricoeur, whose theories
are invoked to support the 'interpretive turn' in the social sciences or the herme-
neutic critique of positivism, to explicate a broader conception – the combination
of interpretation and scientific explanation – in the tradition of Vico, Dilthey and
Weber. Explanation for Ricoeur arises from the human scientific tradition, and
this does not imply the imposition of a positivistic model of science on to social
and cultural phenomena – meaningful action may then become an object of science.

Ricoeur recognizes three movements: the first is prior understanding (*vestehen*) by participating in the human world; the second, interpretation and reflection on that life-world; and the third, an explanation from within the human science tradition rather than the imposition of a positivistic model (Ricoeur 1981:218 cited in Morris 1997:334). Morris claims that most anthropologists have tended to occupy the middle ground between natural sciences and humanities – combining interpretation and scientific explanation; he argues that science is a creative and imaginative representation of reality, and citing the views of both Aristotle and Kant, points out that science is not the only form of knowledge.

Morris advocates realism – not as a theory of knowledge or of truth, but of *being* which forms the basis of common sense. Citing Popper, he argues that science is an attempt to go beyond the world of ordinary experience – seeking to explain the everyday world by reference to hidden worlds similar to religion and art. Defending the classical definition of truth – as the agreement of knowledge with its object (not as an isomorphic reflection of the natural or social world) – Morris argues that knowledge is the search for truth, and that this is a form of scientific cognition distinct from thinking or philosophy, which goes beyond what is known in the search for meaning. Scientific thought does involve the imagination; but science also includes the critical testing of evidence for a particular theory. Most anthropologists, he argues, in favouring a middle position between interpretation and scientific explanation unite the Enlightenment and Romantic traditions (1997: 335). However, the problem is that, even if anthropological knowledge consists of different kinds of knowledge, it is still largely shaped by rationalism, and this is a great handicap in a study of magic. In the social sciences theoretical viewpoints are largely based on the philosophical tradition of rationalism, which places great emphasis on deductive or inductive reasoning and does not accept revelation or 'spiritual experiences' as a source of genuine knowledge. A new framework is required to include magic as another form of consciousness in this scientific understanding of a given life-world or culture.

Such a means – a new mapping of knowledges – has been created by Geoffrey Samuel and is outlined in his book *Mind, Body and Culture* (1990). Samuel goes beyond a phenomenological suspension of the categories of positivistic science, and points out that it is no longer possible to see natural science as the gradual acquisition of objective knowledge in contrast to what he sees as a largely relativistic anthropology. Samuel seeks a more inclusive scientific framework for social and cultural anthropology. The idea of changes in scientific views was first noted by Thomas Kuhn (1970), who suggested that scientific revolutions were the end result of the limitations of previous theoretical frameworks. New theoretical frameworks or paradigms came to replace ones that could no longer explain the world. Samuel argues that the old idea of all branches of knowledge growing out of the trunk of a single tree could be replaced by a new 'paradigm', which is more analogous to the many-centred, multiply interconnected underground network of a rhizome (1990:2). Samuel calls this the 'multimodal framework' or MMF. Samuel developed the MMF as a scientific map with which to examine his data on Tibetan shamanism; and the value of the MMF is that it offers a scientific understanding of the other-world rather than denying its existence. The MMF aims to provide description of the context of multiple ways of knowing in social and cultural anthropology; but above all it seeks to overcome problems associated with dualistic relationships between individual and society, and the theoretical dichotomies between mind and body and anthropology and biology. It also has the ability to go beyond a rationalistic position, such as that adopted by Luhrmann in her study of Western magicians (1989), in making sense of magical modes of operating associated with the otherworld.

Samuel sees the MMF as scientific rather than humanistic and interpretative, a theoretical framework for a natural science of society that provides a language for examining what has been termed informal or non-scientific knowledge – a know-ledge not specifically contained in the mind, but a 'patterning of mind and body as a totality' (1990:6). Samuel views the structures of meaning and feeling in which and through which we live, not as Geertzian 'webs of significance', but by what he terms the 'social manifold', which he describes as patterns formed by the current of a vast stream or river that is directed by the flow of time. At any point in this metaphorical river a cross-section can be drawn to reveal a two-dimensional structure across the flow. The currents are the environments within which human beings move – a field of forces within which activity takes place, and which he likens to Durkheim's 'social facts'. The 'substance' within which this flow takes place, is like 'relatedness' or 'connectedness', and connects human beings, other animals, plants and the environment. Further structure may be imposed on the social manifold, for example, by two theoretical approaches currently employed by the social sciences: individualist and holistic, which are viewed as currents rather than totalizing theories. The process of bringing together what were seen previously to be opposing theories involves not a simple replacing of one descrip-tion by another but a new *attitude,* which Samuel likens to the change in perception adopted by physicists after the introduction of relativity theory. Thus the MMF is a framework for describing human social life in terms of informal as well as formal knowledge (which is based on the informal but crystallized into codes and organization). Informal knowledge rejects the mind–body dichotomy of Cartesian philosophy and sees both mind and body as parallel aspects of a total system. Thus 'truth' is viewed as part of the social functioning of ideas; it cannot be evaluated simply in terms of the currently accepted body of formal knowledge, but is now seen through a variety of frameworks and theories (1990:19).

Notes

1. The differences between 'ordinary consciousness' and altered states of consciousness has been examined by Charles T. Tart (1992[1975]).
2. The otherworld is also called the underworld, the land of the spirits, the world of fairie, or the realm of the ancestors. In high magic it is often referred to as Amenti.
3. The verb 'worship' implies a hierarchical relationship between believer and deity. In contemporary magical practices the magician embodies the forces of the cosmos, represented by deity.
4. In shamanic traditions there are usually three dimensions: upper, middle and lower, although there may be as many as nine worlds, as in Scandinavian mythologies, where they are organized around the world tree Yggdrasil. In most Christian European traditions there are usually three areas: an upper part for blessed souls; an underworld for the damned; and fairyland, an alternative pagan world. The Greeks had Elysium, the pleasant fields in the west for the souls of the blessed, and Tartarus, a place of torment for the damned, although the realm of the dead was generally a neutral underworld governed by Hades (Jones 1996:336).
5. The division of subjective experience into kinaesthetic, aural and visual has been suggested by practitioners of NLP (Neuro Linguistic Programming) (O'Connor and McDermott 1996:xi).
6. 'The self' has a history – the development of the modern idea of what it means to be a human agent is brilliantly demonstrated by Charles Taylor (1989).
7. Except in the case of chaos magick, which emphasizes techniques for changing consciousness and is specifically anti-spiritual.
8. In the case of the Hindu ascetics studied by Obeyesekere, psychological analysis reveals painful experiences that are blocked and symbolized by matted hair. Obeyesekere says that this symbol would cease to exist if individuals did not create it each time by personal anguish. He bases his examination on the *kundalini*. This is a form of yoga practice that focuses psychic energy along the spinal column. In a possession trance the magnetism of the god infuses up the body of the Hindu ascetic priestess through her spinal column. Obeyesekere explains that the matted hair of the ascetic is the god's sublimated penis – a fusion of the symptom and symbol. Eroticism is sublimated, idealized and indirectly expressed (1981:9).
9. The magician put on ceremonial robes and drew a circle on the ground with a consecrated sword to mark out a field of concentrated power that no demon could cross. Cohn notes that at the heart of ritual magic was the belief in the irresistible power of certain divine words that were built into prayers to compel the obedience of demons (1993[1975]:108–10).

10. A term suggested by Ken Rees.
11. Malinowski noted in his study of Trobriand coral gardens that spells associated with fertility and the growth of yams formed a magical system that could not be understood by the ordinary criteria of grammar, logic and consistency. He noted that most magic was chanted in a form of sing-song, and that it was profoundly different from ordinary utterances. The words act because they are seen to be primeval, handed down from magician to magician. Malinowski claimed that magical spells formed a language *sui generis* that could be explained by Frazer's principle of sympathy. The right use of a name had a mystical power that 'transcends the mere utilitarian convenience of such words in communication from man to man' (1978[1935]:233).
12. Both Favret-Saada and Stoller took apprenticeships to understand magical language as a symbolic system. Stoller's apprenticeship to Songhay sorcerers involved memorizing magical incantations that were imbued with power from the heavens and were passed down through the ancestors. These were believed to enter into substances and enable the sorcerer to transform himself into various objects (1987:88).
13. A distinction between magic and religion is unhelpful in a study of magic such as this because contemporary practice includes veneration and control of otherworldly forces. I prefer to take a historical view of magic and religion seeing them both as originating in a pan-human animistic shamanic experience that has been interpreted according to specific situation and culture.
14. Rees makes a distinction between magical space constructed for working magic such as spells and the celebratory magic of the eight rituals of the witchcraft year.
15. If the magician does not feel more empowered, then Rees claims that something is wrong: there is either a 'leak in the hermetic seal' or 'vampiric drainage'.

High Magic: The Divine Spark Within

Magical practice is essentially concerned with a sustained and regular interaction with otherworldly reality, and a magical training helps a magician construct an alternative framework for experiencing that reality. The main emphasis in this chapter is on my training as a high magician. My participant observation was concerned with the elementary stages of a high magic apprenticeship, and it involved what Tanya Luhrmann has termed the 'interpretive drift' – the slow shift of interpreting events, making sense of experiences and responding to the world that a magician makes in learning to make sense of magic (1989:12). From her fieldwork amongst magicians in London, Luhrmann observes that the very process of learning to be a magician elicits systematic changes in the way that events are interpreted (1989:115). Neophytes read about the practice, had 'rich phenomeno-logical experience', talked to other practitioners, and picked up intellectual habits that 'made the magic seem sensible and realistic'.[1] Events were identified as significant, connections between events were made and the new knowledge put the events into context (1989:12). In this chapter I describe the process involved in becoming involved in a systematic magical training as an insider. The process is similar to more conventional fieldwork, which involves learning the ways and mores of a new culture, struggling with a new language, and starting to make sense of the complexities of everyday life. My experiences revealed my own untrained mind before the formal structuring of worldview that must take place as part of the creation of a magical perceptual framework. My journey, like every magician's journey, was largely internal. The language I learnt was esoteric, and my magical training was conducted through meditation interspersed with rituals and meetings with groups of magicians. Thus, my emphasis, contrary to Luhrmann, will not be on how magicians overcome scepticism – people who decide to become magicians are not usually sceptical of magic in the first place – but on seeing the process of becoming engaged in magical practice as learning the language of another mode of reality.

High magic techniques focus initially on training the mind of the magician, mainly through meditation on the Tree of Life glyph (see Figure 1). It is said that it is through the mind that the magician experiences God. In fact the universe has been described as a thought form 'projected from the mind of God' (Fortune

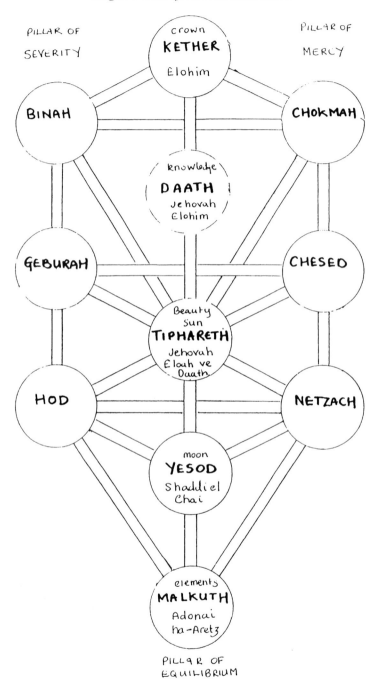

Figure 1. Tree of Life glyph with named spheres. This was drawn by me as part of my homework with the occult school and shows the focus on the middle pillar, the Pillar of Equilibrium

1987b:17). However, to claim that high magic concerns only the mind is a gross misrepresentation of a practice that must engage all bodily experience – the microcosm of the body as a totality. The emphasis on mind reflects Western cultural preoccupations with rationality and intellect, and to a degree it also reveals the difficulties that many Westerners face in experiencing alternative states of consciousness. The main aim of the magical work was the exploration of the self as 'divine spark' in the 're-union with godhead', a process that was viewed as a form of spiritual evolution – the self evolving spiritually through a succession of lifetimes.

Training of a High Magician: Learning a Magical Language

My initial contact with a high magic Order[2] came through an advertisement in an occult magazine. I contacted the organizer, duly filled in an application for membership,[3] and arranged to attend my first monthly study session of the Kabbalah. The headquarters of the magical lodge is a semi-detached house in a respectable leafy suburb of London. The lodge is well-painted and has a neat front garden. I had arrived rather early, so I parked my car and watched for other magicians to arrive. A couple, both wearing black leather and a profusion of esoteric jewellry and chains, looked likely candidates; but they walked straight past the house. A very respectable young man, who looked middle-aged, walked up to the front door and was let in. I plucked up courage to knock at the door. The respectable young man, who was wearing a blue-knitted pullover, answered the door and welcomed me inside. Just as I was taking my shoes off, as requested, there was another knock on the door. The black leather couple entered. They must have walked around the block before deciding to join the meeting.

We walked into the front room, which had its curtains closed. It looked like a very ordinary, rather old-fashioned, front room. On the walls were an Egyptian papyrus painting and an Egyptian bas relief. There were other Egyptian artefacts around an otherwise unremarkable room. The only feature that marked it as rather special was a small occasional table in what I realized later was the east, covered in a purple cloth, which served as an altar. Here Anubis, the jackal-headed Egyptian god of the Underworld, sat alongside statues of Bast (a cat goddess) and the sphinx, and a candle. The 'Hierophant' tarot card, held on a pedestal, formed the main focus. On either side were the two pillars of Joachim and Boas, each inscribed with a pentagram, which is a five-pointed star and represents the four archangels and spirit to high magicians. In front of the altar was a large chart of the Tree of Life, which displayed its tarot card connections (Figure 3). The meeting started at 8 p.m. and was to end at 10 p.m.

There were eleven of us, excluding the lodge officials Eva and Bernard. Both Eva and Bernard were in their early sixties, and looked like any other perfectly ordinary retired couple. Eva, who wrote booklets on astrology, gemstones, and colour in connection with the Kabbalah, was a healer. She led the meeting, and Bernard, who also wrote booklets on similar issues in connection with the Kabbalah, offered advice and comments from his armchair on the other side of the hearth. I discovered that the middle-aged young man, whose name was Paul, had spent two hours travelling from the other side of London. Christine was middle-aged, very vocal but rather rambly in the way she spoke. Mira was a large black woman who was also new and who did not say much. There was a tall black youth who did not say anything either. A stately Asian in his fifties sat at the back and appeared very knowledgeable. Two other women, one of whom had also travelled a long way, were in their early twenties and spoke a lot about their experiences. Another young man seemed to have read a lot about the Kabbalah and asked a lot of questions. The black leather couple, Rebeccah and Dave, were obviously very much concerned with the practical aspects of magic and worked as a couple. Dave was unemployed and seemed 'spaced out' most of the time, while Rebeccah was a research biologist.

Eve started the meeting and explained that this was an 'occasional lodge' to learn the basics of the Tree of Life. We were told that the Tree was a living being that we could learn through the tarot, and that studying it involved a struggle between body and spirit. It was explained that the Kabbalah was Egyptian, not Hebrew,[4] and that the tarot was also Egyptian, orginating from murals of a priest's initiation. Eva started the practical work with an invocation to 'develop the priest within'. After this each person had an opportunity to talk about their month since the last meeting. Those who were experienced used the language of 'tarot-speak' (interpreting their experiences using the symbolism of the tarot and the Kabbalah). I, and I suspect most of the other new members, found this totally incomprehensible. Occasionally either Eva or Bernard would explain a particular point on the Tree of Life glyph that was before us. It soon became obvious that this was a coded type of therapy session where people had time to speak about themselves and their lives and from which they could come to see some coherent pattern using the symbolism of the Kabbalah. Everyone in the room politely listened, and the more knowledgeable made helpful comments. Few personal details were mentioned, but people said where they were on a particular path and what effect that is was having – 'Well, I started this month off in Hod, but moved on to Malkuth, while the Empress was an important factor here.'

After we had all had our say (the other new members and I had struggled to explain ourselves using the totally confusing symbols before us, and had been given a lot of help and encouragement by those who were knowledgeable) it was time for a meditation on the Hierophant tarot card. The Tree of Life glyph was removed, the light dimmed, and we were asked to open ourselves to whatever came through. I knew absolutely nothing about the Hierophant, I was completely devoid of any symbolic meanings or imagery to use, so I looked at the card to try

Figure 2. The Hierophant tarot card. Taken from the Waite-Coleman tarot deck first issued in 1910

to get some information – a stern-faced god-like person wearing a bishop's mitre was seated on a throne with two church officials kneeling before him. What did it mean? I closed my eyes and 'went between the pillars,' as instructed. I experienced a force that felt benign and loving and peaceful – it was just a feeling. We all had a turn to describe our experiences, which, as it turned out, were very varied. Christine spoke about her sense of fear at looking at the Hierophant, and how she could not look directly at him, but only out of the corner of her eye. Comments and helpful suggestions were made to help Christine understand why she was feeling this way. It was pointed out by Bernard that the Hierophant was the guardian of the Abyss – the space between lower and higher forms of consciousness. A couple of people said that they had not experienced anything. There were assurances that this was perfectly normal, and they were told not to worry, because sometimes this happened. Someone else had fallen asleep, and had to be woken up. Paul spoke of the importance of the hierophant's priestly role of guardian of the threshold between the lower part and upper part of the Tree. He experienced the Hierophant as a form of initiation – his life was in the process of change – and he could not make sense of it at the present. In time it would become clear, he just had to wait.

At the end of the session we all sipped tea from bone china cups, passed around the biscuits, and discussed the relative merits of various books, groups and individual approaches to the study of magic; it was also a time for general occult gossip. A great deal of time was devoted to discussing books on the subject, and the more experienced members gave advice freely to the newer members and spoke of their own experiences with certain texts. Paul recommended couple of books on the Kabbalah to me and one woman gave me a copy of some rituals. In due course Christine's husband called to collect her, and it seemed to be the right time for us all to leave.

During the intervening month, all the new members of my Kabbalistic study group were expected to read extensively about the Kabbalah. The aim of the reading was to increase knowledge and thereby their personal involvement with the glyph – it had to become second nature to us. So much so, that we had to aim towards thinking in Kabbalistic symbolism all the time. Bernard explained the growth of his garden, his work weeding it, the personalities of the birds that came to feed in it in Kabbalistic terms. A car crash he witnessed was explained by opposing forces on the Tree of Life. In short, life is an expression of forces or energies, and these various forces have been systematically classified in the Tree of Life glyph. The Kabbalistic magician has to learn where to locate the forces on her or his inner cosmological 'map' of the otherworld. During my fieldwork the feeling of struggling with the comprehension of a new language was never clearer than when I was starting to learn the Kabbalah. To begin with it felt like a completely foreign vocabulary that was full of esoteric symbols and meanings. Slowly, very slowly, I began to understand the symbols and the way they related to my life – my cosmological framework was being slowly shifted.

I felt more relaxed at my second study meeting – I was finding that the Tree of Life was becoming more familiar as I tried to apply it to my life. I had been helped in this by a wiccan high priest I had gone to interview for the research, who had an extensive knowledge of the Kabbalah. He had spent hours with me patiently answering my questions, explaining the complexities of the different interpretations of its meaning, and drawing complicated diagrams of the attributes of each sephirah (Figures 3 and 4).

In the end, perhaps almost in desperation, he had told me to strip away anything which was not personal, and had encouraged me to meditate on the sephiroth to find out what they meant to me rather than what they were said to represent in a text. This indicated an important move inwards away from 'received meanings' – I was learning to relate the language to my own experience.

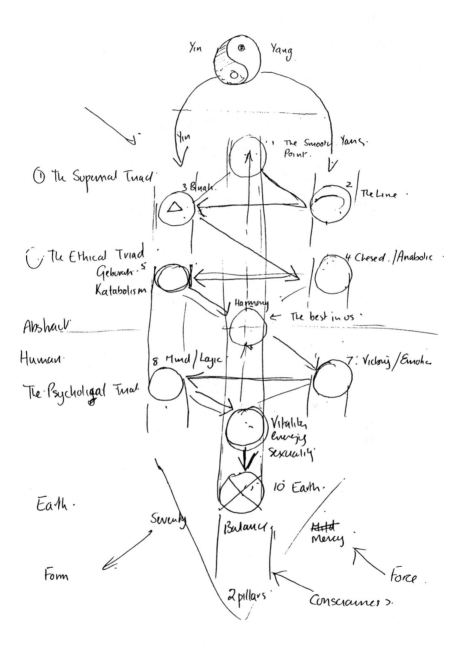

Figure 3. Diagram showing the 10 spheres of the Tree of Life and their relationships with one another and to the human microcosm as drawn for me by a wiccan priest

That which is above is as that which is below, after another manner.

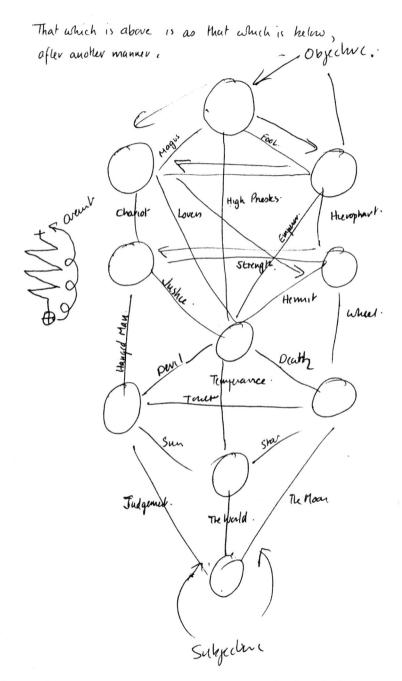

Figure 4. Diagram showing the paths connecting the 10 spheres of the Tree of Life. Each path relates to a Tarot card and links one sphere to the others. The small figure on the left shows the flow of energy

I reported to the group that I had been stumbling around getting to know the Tree, and that I was working on the sephiroth relating my own experience to their energies. There were a lot of murmurings of support and help: 'Yes, this was a good thing to do.' Christine suggested using elements that were already known to me, such as the zodiac. We went round the room, and this week there were another two new members, a young African-Caribbean man who did not say much and a young Scottish man who was interested in the Merlin tarot and kept asking when and how objective experience becomes subjective. He was very keen to detach himself from his emotions so that he could analyse them.

We settled down for another visualization on the Hierophant, the same tarot card being used for two or three months running so that we could develop a deep relationship with its inner meanings; and this was meant to increase our involvement with it. Each meditation on it was supposed to increase awareness and depth of experience. However, this time I experienced difficulties with the visualization. I felt totally resistant, and my legs and arms were twitching – I did not want to participate in the trance work. I attributed this to tiredness and a growing antagonism towards the work. When it came to my turn to describe my experiences I explained that I felt resistant to the Hierophant. I was told that this was sometimes a common response and that it did not matter. I should try again on another occasion. Others had good visualizations. Christine described entering a vast temple with coloured tiles on the floor and mosaics on the walls. She said that she felt overawed as she watched herself walking along in a very special place. Rebeccah felt that the Hierophant; was her guide on the Tree – her teacher of the paths. Paul became very involved with the two kneeling monk figures on the Hierophant card. He felt that they had a message for him as he knelt beside them and looked up at the powerful seated figure of the Hierophant; but it was too early to tell what it all meant.

This month there was a short talk on ritual before we performed a ritual. It was explained by Eva that we did rituals to ask for things from the 'Higher Things That Be' for other people, not ourselves. The one exception was to ask for protection, because a 'vulnerable occultist was no good to anyone'. Ours was the 'right-hand path' for help and healing. We were not to impose our will on anyone. If we asked for healing for someone it had to be with their permission. If we asked for wisdom, we had to ask that wisdom be made available to someone, should they wish to use it. Eva also said that she was reporting back from another member, Graham, who wished us to know that his ritual work had been rewarded – 'although we must not keep looking to the ends, it was encouraging to sometimes see results', she said. The first of Graham's rituals was for Russia, and Eva said that Graham had felt that his work had helped to overthrow the coup. The second was that he had found himself in the position of having to find a home for an elderly person. Initially he had no idea about how to go about this, but two days later a home 'had materialized'.

The group ritual on this occasion was performed for what is now the former Yugoslavia. We were told to be very specific in what we asked for, and we had to be very careful how we asked, because 'the Cosmic Forces are amoral' – they can be used for good or evil. So we had to be careful to ask for an end to unjust oppression – 'some oppression may be justified, as we do not know what has happened in previous aeons'.

All the chairs were cleared to the edges of the room, and a circle of protection was created to keep out evil and demons (see Kieckhefer 1993:6; Cohn 1993:102). The performance of 'The Lesser Banishing Ritual of the Pentagram' is essentially a means of making a focus of magical power through the invocation of angels; it also concerns the purification of the mind (Regardie1991a[1938]:96–7). Those who knew the movements of forming pentagrams in the air in the appropriate directions got up and stood in the centre of the room, while the newcomers gathered in a corner to watch and take notes. I was handed a sheet of directions:

The Lesser Banishing Ritual of the Pentagram

'I invoke the Mighty Archangel Raphael to protect me from all evil and aggression approaching from the east' (visualize Golden Yellow with a dash of Amethyst).

Turn to the other Quarters in turn:
South: Michael – Red/Emerald Green.
West: Gabriel – Blue/Orange.
North: Auriel – Olive Green/Russet/Citrine/dash of Black.

Back to the East.

Take to hand the mighty sword:
'In Thy Name, O Ain, I take to hand the mighty sword as a protection against all evil and aggression.'

Put sword down and turn in full circle as a protection – making a circle of protection around you with the light. Sheath sword. Make a Pentagram of yourself and say:

'Before me Raphael'
'Behind me Gabriel'
'On my right hand, Michael'
'On my left hand, Auriel'
'Before me flames the Pentagram'

These directions utilized a sword (which is frequently used for casting circles), but the magicians in front of me performed the operation with pointed fingers instead. The magician Israel Regardie claims that the overall result of performing these actions is that gradually 'coarser elements are ejected from the sphere of sensation'. Other particles, more 'sensitive and refined', of a 'higher grade of spiritual substance', are thought to be attracted to the personal sphere and 'become infused into the character and nature of the physical and psychological constitution . . .'. A purification is believed to take place, enabling the influence of the 'higher Genius to penetrate the refined and porous brain to diffuse throughout the personality' (Regardie 1991a[1938]:96–7).

This was followed by a performance of the 'Kabbalistic Cross':

The Kabbalistic Cross

Visualize a ball of white light above the head.

1. To thee, O Ain – Touch Forehead
2. Be the Kingdom – Touch Navel
3. The Power – Left Shoulder
4. The Glory – Right Shoulder.
5. Clasp left hand over right hand in front of you.
6. Say 'Ages unto Ages' – Stand and meet light coming down.
7. With legs slightly apart make a pentagram:
   Face east (the position of the rising sun and first light).

The main reason for performing the Kabbalistic Cross ritual is for the magician to get in contact with her/his 'higher self', which is visualized as white light, and drawn down from above the head. Regular practice is said to induce a recognition of the higher transcendental self, which bridges the conscious and the unconscious mind (Regardie 1991a[1938]).

In the creation of sacred space we were able to contact our higher selves and we focused on the object of the ritual – the sending of wisdom to the former Yugoslavia. We asked for 'wisdom from the west' so that the people in the former Yugoslavia be freed from unjust forces and that wisdom be made available, should they choose to take it, for a peaceful solution.

At the end of the ritual, when the sacred circle had been closed, people rearranged the chairs and a sense of everyday reality returned. Nobody spoke much about the ritual or their experiences, and the talk centred on more mundane matters before the meeting broke up. I was left feeling rather confused about the mechanics of what we had done – would I ever learn the complicated hand and arm movements, and would the Kabbalah ever make sense?

However, by the time of the third meeting I felt that I was at last beginning to make sense of the strange language of the Kabbalah. I had been working consistently on the glyph in the intervening period, and was beginning to internalize it and relate it to my life. As we went round the group, with each person having the opportunity to speak, I felt as though I was understanding the language at last, and I felt that I could contribute a little – I was at the stage of learning the common nouns and asking basic questions in a foreign language, such as 'Excuse me, can you tell me the way to the station'. I said that I was working my way around the Tree and was beginning to understand that I was feeling centred in Yesod and Netzach, while my work was located largely in Hod. Paul suggested that I was on the 'path of Strength', and I realized that I would be expected to go home and ponder on this suggestion and look up the various attributes of this specific path.

The meditation on the Hierophant was very different this time. I was concerned that I would feel resistant again, but I was determined to experience something, so I 'marched' between the pillars before any thought of resisting could stop me. An extract from my field notes ran like this:

> I was conscious of the Hierophant there as a presence, but I did not look at him as I was too busy looking at a great circular abyss which was wide, deep, and was giving off a hazy white mist. I was aware of the presence of the Hierophant but he seemed benign and helpful. I wandered around the abyss not knowing what to do, and before I knew it, I had fallen into the depths and was going down and down into the dark of the Tree of Knowledge – down and down until I came to the very bottom. The Hierophant was with me as I came to face myself. I looked at myself and I was covered in black veils. The Hierophant helped me to peel off the layers, layer after layer, until I could face myself and I flowed into the whole Tree and became a part of it.

This experience came as quite a shock to me and I felt a bit embarrassed about it. I was not sure if it was the type of experience that was 'required'. It seemed very different to other people's accounts and was obviously a mixture of various 'descent myths', in particular Inanna's journey to the Underworld. Their experiences seemed very contained and orderly in comparison with mine. A couple of people had described walking through halls and temples, and gave exact details of different people they had met, colours of stained glass windows they had seen, significant animals they had passed on the way, and how they had felt in relation to their lives at the present time. I felt rather embarrassed about my chaotic flowings and, as I was not directly asked to describe my experience, I decided to keep it to myself. At this stage in my fieldwork I felt very apprehensive about talking about my magical experiences. If I had had this visualization towards the end of my research I would probably have felt more confident of opening up a discussion on the subject. My experience does reveal different magical approaches to the otherworld. High magic in particular is very structured, it has a very orderly approach; its principal teachings frequently use images, which are often located within a specific mythos, but invariably include figures in robes and cloaks, tapestries, temples, altars, curtains revealing doors (to another level of consciousness), animal and angel guides, planetary symbols, ancient heroes and heroines, journeys to healers, teachers, elders or 'the self'. It also employs an archaic language to create a sense of timelessness, depth and tradition.

In a practical manual of guided pathworkings, Dolores Ashcroft-Nowicki takes the reader into the temple of Malkuth as preparation for the Thirty-Second path of the Kabbalah:

Awareness of our normal surroundings slips away gradually and in its place the temple of Malkuth grows around us. It is square in shape with a floor of black and white tiles that feel cool to our sandalled feet. The north wall is on our left, the west behind us, the south to our right. Set into these three walls are circular windows of richly stained glass, each a representation of the Holy Creature of that quarter. In the north it is a winged Bull, set in a circle of golden wheat and scarlet poppies. Behind us in the west, an Eagle soars into the sun through a sky of brilliant blue. In the south a winged Lion stands guard surrounded by flames.

Before us in the eastern wall are three heavy oaken doors; they have no handles and no locks. In front of the doors stand two pillars reaching from floor to ceiling. As we face them, the pillar on our left is made of ebony, that on the right of silver. Both are topped with a capital bearing a carved and gilded pomegranate. In the middle of the temple stands the altar, a double cube of black polished wood. Covering it is a cloth of handwoven linen on which are scattered ears of wheat.

On the altar is a bowl of deep blue crystal in which burns a light. This light is to be found on every altar of the Mysteries no matter in which tradition it is set. Unless it be lit, no temple is truly contacted. The reflection of this light is carried within the heart on every journey, a protection for those who travel and a symbol of light for those we meet in the inner worlds. Above the altar hangs a bronze lamp that burns a scented oil filling the temple with a subtle fragrance.

Behind the altar stands the figure of Sandalphon, the archangel of Malkuth. He appears as a young man with dark curling hair in which are twined clusters of grapes and vine leaves. His eyes hold a wisdom gained while the earth was still young, and a sadness that this is no longer so. His robes are a mixture of russet red, gold, and apple green. The air about him shimmers with a faint radiance, an aura of power that belies his gentle glance.

We stand before the altar, making ourselves ready for the journey ahead. Then Sandalphon moves to the centre door and draws in the air before it a five pointed star. It hangs flaming for a few moments, then fades; as it does so over the door forms a curtain depicting the tarot card of The World. It glows brighter, then becomes a three-dimensional door to the thirty-second path. We walk between the pillars, the dancer hangs motionless within her wreath of leaves, and we step forward into a swirl of colour (1983:23,24).

Several points need emphasis. Firstly, the shift in consciousness as we enter the temple – we are on a journey to the otherworld. Changes in consciousness are often symbolized by doors as gateways to different levels of awareness. Secondly, the temple is structured on Masonic imagery: black and white floor tiles and black and silver pillars are an important conceptual device representing polarity and balance – the awareness of the whole. This makes an interesting comparision with witchcraft rituals, which are typically conducted outside 'in nature'. Thirdly, the internal link to a higher self or divinity, which is frequently represented by the notion of wisdom – this is a journey undertaken with a purpose: the finding of the true self and its connection with the cosmos.

Deeper In: An Apprenticeship in Magic

Christine had been attending the monthly sessions for twelve years, but I felt for the purposes of this research that I needed a more intensive and structured approach to learning the Kabbalah, and so, on the advice of a female magician, I applied for membership of a well-known occult school offering a correspondence course that included a more intensive and structured approach to learning the Tree of Life. I completed a detailed application form, which requested information on my religious upbringing, education, job or profession, marital status, and esoteric group member-ship, on whether I had ever taken drugs or had a nervous breakdown, and finally on what were my reasons for wanting to join. In this last section I made it clear that I was conducting anthropological research. I also had to complete an examina-tion on my knowledge of symbolism, mythology, ritual and meditation. Had I had any psychic experiences? The results of this examination were marked both conventionally and 'psychically', and took some time to reach me.

I eventually learnt that I had been accepted. My magical supervisor, who was called Pearl, wrote to me to introduce herself, and my apprenticeship started on a daily regular basis. The early lessons were concerned with developing memory skills, relaxation, and learning the basics of magical practice, including how to keep a magical record of meditations. I had daily assignments, and these were sent off monthly to Pearl, who commented on them. An important part of the preliminary lessons was concerned with creative visualization. I had to practice 'seeing routes' in my mind. A record from my magical diary read:

> I have been practising 'seeing' routes. I found the best time was in bed before I went to sleep. I started off going down the road, trying to remember exactly each significant detail as if I was giving directions. Then I practised the route to the nearest town; this was much more difficult than I had imagined, but after practice, I could remember clearly. Then I tried the journey which I take my daughter to school – this is some three miles. To begin with it was extremely difficult but the next time I took her I made a conscious effort to remember and this made all the difference.

Pearl commented that she thought that the lessons on observation were a waste of time when she first did them, but that she had learnt since that they were 'SO important':

> . . . I remember that I thought that it was a waste of time when I did it. I find even now, that I will suddenly notice something on the way to work, and realised that this lesson is still happening for me. You are probably aware that it is recommended for practitioners of the occult to review the day in a similar way to how you have been seeing the routes in this exercise. Try to walk very step of your journey rather than hovering above the pavement, and notice the mundane as well as the exciting and interesting things.

I had to remember the first film I saw, and also my first day at school. In another exercise, I had to think of the room I had had as a child:

> This exercise brought back some difficult memories and some surprises . . . remembering the rooms in our house recaptured how I felt as a child. The amazing thing was how much I remembered – the wallpaper, the colour of the carpets brought back what life was like then. In particular I remember in detail the wallpaper in my room before I moved to a larger bedroom after my grandfather had died. The wallpaper was pale mauve and was arranged in boxes, each with a flower inside (I think they were primulas). I also remember the rowan tree outside the window; I used to lie in bed looking at the tree, it was a very special tree . . .

In another exercise I had to think of a picture in a friend's house, to recall as much detail as possible. I tried to visualize Manet's 'A Bar at the Folies-Bergères'. Pearl advised me to go into the picture and see the rest of the bar as the barmaid:

> When you have done that, feel what she is feeling and think what she is thinking. Manet knew all of this information when he painted the picture, so the information is accessible to you as well. I sometimes use this as a form of relaxation (depending on the picture!). Over the years, I have learned that a short cut is to be had by focusing on the smell in the painting, and the rest seems to follow quite easily. All you have to do is remember the painting clearly . . .

This is practical training for the mind in preparation for pathworking and voyages to the otherworld. It is a Mnemonic system, which prepares the occult student for advanced magical work. Mnemonics had a long history in the Middle Ages, and became fashionable among Hermeticists as a method of printing basic or archetypal images on the memory. Yates argues that the Hermetic experience of reflecting the universe in the mind is at the root of Renaissance magic. This is done by using magical or talismanic images as memory-images by which the magus hoped to acquire universal knowledge, and also powers, obtaining through the magical organization of the imagination tuned to the powers of the cosmos (Yates 1991:192).

The emphasis is on mapping and accessing the hidden or unconscious aspects of mind as a rich otherworldly resource. The otherworld may be brought through to ordinary consciousness to develop and empower the magical self. But first the magical self has to be found. My early lessons focused on a questioning of the self. An entry in my diary record shows:

Thursday 16 January 1992
Bedroom 9-00pm – 9-15pm
Relaxation: OK – did Yoga exercises first   Breathing: good
Meditation Subject: 'Behind my conscious self is my secret self'

Behind my conscious self – what is that? My conscious self – what I am aware of. Conscious implies unconscious. Is my secret self unconscious? What is my secret self? Do I have a secret self and is it 'behind' me? Is my secret self my 'inner undiscovered self'? Is it undiscovered because it is unconscious?

My inner self is partly conscious and partly unconscious – the unconscious parts are reaching out to the conscious parts WHEN I LET THEM – i.e. when I give myself time; sometimes they emerge in dreams, often they emerge as profound realisations from 'pathworking' the Kabbalah. Is my inner self secretive? Do I have anything to hide?

Pearl, my magical supervisor, explained that: 'Your secret self is your unconscious and higher selves. Put them all together and there you have a whole beautiful you, containing the Divine Spark. So why would it be stupid to expose it?'

Another meditation read:

Friday 17 January 1992
Bedroom 10.30pm – 10.45pm
Relaxation: OK – bit of Yoga   Breathing: good
Meditation Subject: 'Behind my conscious self is my secret self'.

Straight into it this time. Accepted the idea that my secret self is my INNER SELF. Felt like a voyage of discovery through my forehead chakra straight down and down inside myself. I saw myself from the inside out – instead of the outside of my arms and legs and body I saw the hollowness of myself inside – I kept falling and falling through this vast vault of a thing which is my body. And as I fell I got smaller and smaller like Alice in Wonderland and I wondered when I was going to stop falling. Eventually I landed at the bottom of myself which was my root chakra (I think). I felt small and very needy – like a child – images of my childhood came back to me as I thought 'what am I doing here?' Then I had a vision of Malkuth as Gaia – the earth slowly revolving and me standing looking on and being aware that this was the start of my journey. Then I felt that I'd had enough and wanted to stop.

Pearl commented that the feeling of getting smaller is usually a sign of early changes in consciousness, and that this was a good sign. She also said that: 'East and West don't mix as both stimulate the same chakra in differing amounts and in different ways. If you do this, it is the same as following two completely different recipes using the same mixing bowl.' She said she tried mixing them once and could still remember how the next day she was sick out of her car window while waiting at some traffic lights. This high magic school differentiated between 'Eastern' and 'Western' methods of raising energy. The eastern tradition of raising 'kundalini' energy is conceptualized as a sleeping snake coiled at the base of the human spinal column. Using the appropriate techniques, this energy can be made to ascend 'channels' either side of the spine up to the head. The resulting experience is termed 'enlightenment', and forms the basis of much Tantric sexual practice. By contrast, the method used by Western magicians is to work through the mind. Energy is visualized as being drawn down from above. Western magicians see this as safer, because the mind is believed to work as a 'safety valve' to 'block' energies that are too powerful for it fully to comprehend. In short, Eastern techniques generate energy that is visualized as flowing upwards from the base of the spine, whereas Western high magicians prefer to mediate the energy through the mind. However, I was to learn later that high-grade magicians certainly do use Tantric practices.

Some meditation led to specific visualizations. A later entry into my magical diary shows how I was beginning to interpret my meditation using 'appropriate' language. I had stopped referring to the Eastern chakra system and was thinking in 'Egyptian' and using Kabbalistic terms:

Sunday 25 October 1992
Bedroom 10.15 – 10.30pm   Relaxation: Relaxed   Breathing: OK
Meditation Subject: Let the White Light of the Divine Spirit Descend

Vibrated God names and felt all my being attuned. The white light of Kether filtered down very slowly, then it turned into white water and slowly trickled down over what seemed parched, dry and arid land. This in turn became the desert . . . I found myself travelling up the Nile, which was the white light, up to two figures of Osiris and Isis; both were seated on enormous thrones and appeared very grand. I felt very small but after a time I felt very attuned to them and very linked to them, although not a word was said.

Other exercises were concerned with developing the imagination using mythology. Stories of the gods and goddesses' adventures are seen to be 'parables of the workings of the cosmic forces' (Fortune 1987b). I had to choose a myth and every day build it up in my imagination, paying particular attention to details. The following week, I had to imagine that I was one of the characters. I chose to be

Persephone, from the Greek myth of Demeter and Persephone, and my diary record read like this:

> I am Persephone. I am sixteen and I am playing with my friends outside my house, actually its the back of the house, in a ploughed field. I am playing a game of 'it' with my friends. You may think we're a bit old to be playing such childish games, but its a lovely sunny evening and we feel full of fun and in a lively mood . . . The air is still, the sun goes behind some dark evergreen trees and suddenly fades, something strange is happening. We all stop laughing, we can all feel it. There is a strange feeling. I lie on the ground still, my heart pounding. I can feel the ground damp and cool under my body. My friends seem distant, I am losing touch.
>
> As I lie on the ground I seem to become at one with the ground, it opens up and takes me in. My friends are gone, I am alone, all alone. I feel my loneliness, the difference is startling, yet I do not feel worried. I feel the lack of sun, I see the dark overwhelm me. I feel as though I'm falling, falling downward, spinning. I'm falling. It seems for ages, years even, but is probably only a second or two. I shut my eyes, yet I'm still falling – this is not a dream. I open my eyes and focus clearly on what seems like the sides of an enormous hole where you can see the different layers of earth and chalk the deeper and deeper you go. I focus hard on the sides of the hole and as I focus it appears to change from earth to sky, the night sky with millions of tiny planets – stars glistening. I shut my eyes – I cannot take any more of this. It is just too much to comprehend – is this down or am I flying up?
>
> I do not feel fear – in fact I don't feel anything – should I feel something? I know I have to go – it is my destiny, my journey. I am undertaking my spirit flight. In fact I have lost my physical body. I know it has gone but it doesn't worry me – nothing worries me any more . . . I feel lighter and lighter – I look down at myself and I'm all shimmery and silvery – still me but different. As I look down I feel that I have stopped 'falling' (maybe I wasn't falling anyway). I glide along what looks like a roadway – but can it be a roadway? Where am I? I am stopped by another shadowy silvery form who asks me my name. I respond "My name is . . .?" For a moment I've forgotten who I am – what I am. I stop and think. I have come from the Upperworld, that's for sure, and I've come on a journey. But what was I and where have I come? I am overcome by a terrible feeling of confusion . . .

This account shows clearly my untrained mind; it contains elements of different traditions – such as the shamanic spirit flight – and it is largely external – told as a story. When I sent it to Pearl she said that the tale was 'beautifully told', but what she really wanted to know was the inner meaning of the myth:

> For example, we all come from our mother Earth and to her we all return. When did you eat the pomegranate? If you are Demeter and your animus is Zeus, how do you deal with the decisions 'he' takes that you don't like? Who is the Hades aspect of you? Who is the Hermes that you send to talk to your Hades? The questions are endless. Now let's look at Psyche and Eros. He was sent to kill her and then thought better of it

and married her instead. It never does to kill your anima. She disobeyed him and he did a bunk with his injured pride. After spending a long time reflecting on her predicament, Psyche decided that she loved him and like Persephone, went to the underworld – except that Psyche was in total despair. If you want to find God, look on the end of your tether! Finally, Venus (the bit of her that had caused the whole mess in the first place) took pity on Psyche, and reunited her with Eros. But the whole point of this myth, is that if she hadn't been human with all of the human failings, she could never have become immortal – which is what happens when your male and female parts get together.

In this way myths function as inner dramas that help the magician negotiate their inner energies according to wider cosmic forces, or, as in this case, the Collective Unconscious as interpreted by Jungian psychology. Pearl asked me to look at the story of Snow White and tell her its inner meaning: 'Remember, the apple (knowledge) put her [Snow White] into the shock of deep sleep and she was awakened by the Kiss of the Prince.' Pearl was working on awakening my animus.

Pearl's response helped me to see that my experience of visualization with Eva and Bernard's Kabbalistic Order was different from the other accounts because I had not related my knowledge of Kabbalah to my own inner energies – in short I had not learnt to use the Kabbalah as a route map for linking myself with the cosmic forces. Likewise, in experiencing the myth of Demeter and Persephone I had not interpreted the myth's inner meanings. I was learning that the magicians' maxim 'as above, so below' had a very intimate significance.

Some time into my training I had the opportunity to meet and talk to Pearl. Our relationship had been solely based on lesson correspondence; but the occult school's annual Conference held in London offered the chance for us to get to know each other personally. I had asked Pearl previously how I would recognize her at the Conference, and she replied that 'we would know each other'. I tried to keep my mind open to 'knowing her' as I walked into the hall. I looked around, but nothing 'happened', so I sat down and listened to one of the talks. A couple in their thirties, who looked as if they had walked right out of the City, sat down beside me. He was dressed in a suit, very smart, and she looked as though she was his secretary, wearing a mini skirt, black tights, stiletto heels, and a lot of make-up. After the talk I walked out to look at the bookstalls and got into conversation with a woman I recognized. I looked up as the secretary walked past, I saw her name label: she was my supervisor! Obviously the psychic connection between us had not material-ized. Seeing her was quite a shock, as she was very different to what I had imagined. However, I introduced myself, and we went into the bar to talk. Pearl told me that I looked much better than my photo (I had had to send in a photograph of myself when I applied for entrance into the occult school), and that she thought I was a 'Green Ray' earthy type. I learnt later that this was a reference to Dion Fortune's teachings about Rays of consciousness. For example, the Christian Violet Ray predominates in the West, while the Orange Ray of the Buddha prevails in the East. The Green Ray is the connecting link between the Red Ray, which concerns the development of the individual, and the Violet Ray, which is associated with Group Minds (Fortune 1987:162). At the time I was unaware of the complexities of these correspondences, and instead took it as a compliment that I was connected to the earth. She asked me how my studies were going, and I explained that I was having difficulty with the Middle Pillar exercises. She told me to put aside all of my belief systems, including ideas about patriarchy: 'We're all the same on the inner', she said. Pearl explained that later in the course we would cover the path of the magician and that we would link Chockmah (a masculine sphere) and Binah (the feminine counterpart).

I asked Pearl, who said that she also had links with the Anglican Church, how she had got involved in the occult school. She said that she first started practising magic when she was at college. At first she was very wary, thinking that the group might be practising black magic, and she had asked her mother, who was an adept in Dion Fortune's group, for advice. Her mother had suggested caution and to 'go and see' what the group were like. Pearl said she need not have worried, because they were not practising black magic, and her involvement with them was the right thing for her to do. It had completely changed her life, and she now had a new husband, a new house and a new car. She met her second husband through the occult school, and they had 'teamed up'.

My training continued, and all the time I was learning more about the essential principles of high magic. Pearl warned that shortly changes would happen to me: something like a combination of the films of the Pink Panther and the Night of the Long Knives.

Meditation and the Kabbalah

The Kabbalah is the cornerstone of much high magical practice. It is composed of symbols that can be seen by the eye and used to concentrate the mind in order to introduce certain thoughts, associated ideas and feelings. As has been mentioned previously, Dion Fortune wrote that the meditation on symbols unfolds what generations of mediation have 'ensouled therein' (1987b:5) – symbols are thus seen to be latent 'power houses' of energies that have existed since the beginnings of time and that can be passed on to present and future magicians as esoteric 'cosmic truths'. Symbols are also seen to be a means of guiding thought out into the unseen and incomprehensible; 'by thinking we create concepts' (Fortune 1987b). Some of my meditation practices concerned visualizing symbols in this way:

High Magic: The Divine Spark Within

4 July 1992
Bedroom 11pm   Relaxation: OK   State of Mind: tense
Meditation Subject: Two triangles – one pointing downwards, one upwards

I went quickly into this one: the bottom triangle was sitting in me as I sat on the floor, the top one was balanced on my shoulders and they met midway. The colours of both triangles seemed to change quite a bit. The bottom one felt very blue most of the time, although it did change to red at one point. The top one was white and then blue. To begin, both triangles were very clearly formed but as I got into the meditation the top one seemed to expand and totally encompass the bottom one. It felt very strong and I was aware of a highly charged force which seemed to encompass me in a feeling of white light. The point where the two triangles met was like it was charged with lightning.

Thought is the start of the meditation process, but it is said that a knowledge of the higher forms of existence must be obtained by a process other than thought. Symbols are said to be a 'stairway of realization' to climb 'where they cannot fly'. Symbols are also 'to the mind what tools are to the hand – an extended application of its powers' (Fortune 1987b:29–31). However, to a new apprentice, these seemed very lofty ideals. Some meditations did not work very well, and a lot of meditation practice was repetitive, boring and appeared to lead nowhere:

5 July 1992
Bedroom 11pm   Relaxation: stiff.   State of Mind: OK
Meditation Subject: Two triangles – one pointing downwards, one upwards

It was difficult to get into the meditation this time. Yesterday's was quite powerful and I couldn't get that image from my mind, I kept thinking of the energy, like a lightning bolt between the two triangles. After the two triangles had merged with one another, I decided to leave it.

One meditation, early on in my apprenticeship, seemed to take on nightmarish qualities:

My forehead was drawing me in – as I went into myself – I found myself in my head – but it was dark and a mass of tunnels – almost like sewers – these changed to an underground railway system with trains going in all directions and escalators in front of me. I didn't know which way to turn and just watched and waited until I knew what to do. After an awful long time it came to me that I should place my forehead on the cold tiled floor – as I did so I felt myself relax and 'descend' into my body . . .

In later meditations I learned to handle 'the force':

> Monday 9th November 1992.　　　10.15 – 10.45
> Bedroom: Relaxed & breathing ok.
> Meditation Subject: 'Thee I invoke Bornless One'.

> This was a powerful meditation, I felt tremendous pressure in my head as I thought about the meditation subject 'Thee I invoke Bornless One'. I connected to deep space and saw Nuit creator of all arching over the night sky of the universe. Small stars glittered and I was drawn into one. Then my head started rocking and I felt I must draw the universe down into me. There was a tremendous explosion of energy which rocked my whole body until I remembered the fountain method of the circulation of force [a technique for circulating energy around the body]. This being done I calmly drew the starry energy of the universe through myself in a relaxed way.
> Realisations: it's extremely important for me to have an image of God and Goddess.

Pearl had told me that I worked 'strongly with the feminine' and that I should 'think more of the masculine', and that I should see Kether with a male face as 'sky energy', which was paradoxically also female too. I think I was proving to be a 'difficult' student – I was asking too many questions. Learning a magical worldview takes a certain amount of acceptance from a student that their supervisor has adequate answers and explanations for their questions. I found that the more questions I asked Pearl the more my progress within the school was hindered. It reached such a low-point that Pearl asked me to 'talk to' Hypatia. Pearl explained that Hypatia was one of the seven teachers of Alexandria, and that I should build a picture of her and draw it into my solar plexus, then let it rise to Kether and down to Malkuth. She asked me to 'talk to her' for three nights in a row. The essence of my meditations went as follows:

> Nectar to the soul! She is the Divine feminine – all that has been ignored, repressed or mis-represented by patriarchy and by patriarchal interpretations. She said that she was there – she was the Middle Pillar – I must connect to her before I can fully experience polarity or balance – she is the redeemer of women.
> Hypatia told me that she was a part of this energy force – a part and yet not a part. She is a paradox – there – yet not there. Within me yet not within me and both.

I realized that if I was going to continue that I would have to accept unquestioningly the framework of the practice. As soon as I did so my relationship with Pearl improved dramatically, and she started encouraging me and telling me how much progress I was making. It was shortly after this that she asked me if I was interested in joining a magical lodge at the London temple. I had had the opportunity to compare my progress with that of another one of Pearl's students at the annual Conference. Although at the same stage of the course as me, he had been invited to various rituals at Pearl's house and had already been asked to join a magic lodge. I realized just how much my attitude and my questions had hindered my magical development in this school.

As I stopped asking questions and meditation practice progressed, I did experience the glyph as inside my body, and it did come to represent a map on which I could label and interpret the experiences that I was gaining through my daily meditations. At a certain stage, I was mentally able to build the complete tree of life in its four manifestations, using the different colour symbolism for each manifestation, guide my breathing, vibrate the god names for each sephirah and focus on a sentence for meditation.

Pathworking

As was mentioned previously, an important part of magical training is the guided meditations called 'pathworkings'. The monthly Kabbalistic study group was based on 'pathworking' the Kabbalah using tarot images from a tarot deck. A flyer for a two-day workshop given by Dolores Ashcroft-Nowicki, a 'pioneer' of this technique, at the College of Psychic Studies, explains that pathworkings are not fantasy games in the mind but ways to explore inner worlds that can be used to 'ease stress, to train the memory and sharpen the mind, to improve performance in sport and help you over interviews, exams and to work out problems', as well as for healing and entertainment: 'The Inner Kingdom of the Mind is a fascinating place, it is the last frontier left to explore and it has no limits in time or space. But learning to use this faculty without allowing it to become addictive means one has to approach it in a disciplined way.' Each sephirah of the Kabbalah is seen to be connected to another sephirah by such pathways. Ashcroft-Nowicki says that these pathways represent the forces and energies of their connecting sephiroth, and it is here that magicians believe an understanding of the unconscious mind is possible. Paths have been described as 'pools' or 'reservoirs of influence' that act as 'termini for the beginning and ending of each path'. They are seen to collect and hold traces and echoes from the paths that enter and leave them, so that each sephirah contains 'particles of influence drawn not only from its fellows, but also from their mutually connective paths' (Ashcroft-Nowicki 1983:11).

Pathworkings are seen to take the magician out into the astral plane – as an 'amorphous mass waiting to receive the imprint of a thought form' – to expand consciousness. The paths on the Tree of Life are equated to the 'creative astral', a fluid world where things 'can, and do dissolve as soon as you take your mind off them'. Ashcroft-Nowicki emphasizes mental discipline, in order to 'hold the images' for periods of time (1983:14). Pathworkings formed the basis of many rituals, workshops and seminars that I attended. Much high ritual practice is concerned with enacting pathworkings as ritual drama. One pathworking, given by Dolores Ashcroft Nowicki at a conference, went like this:

Breathe easily and deeply, establish your own pattern. Begin to relax. Build a shell – warm, soft, darkness – back inside your mother's womb – warm, soft, safe. You hear the faint sound of your own heartbeat. Allow outside noises to go through you. All that matters is that you are drifting gently. You begin to spin, tumbling head over heels, down and down – until you come to rest on something soft and springy.

Your inner eye begins to open – you are lying on a green hillside. [This is **Malkuth** – at the base of the Tree of Life.] Below you is a valley – you can see a small city. . . . Dawn has just broken and you feel refreshed. Get to your feet, walk down to the city. The gates of the city swing open as you approach. It's rather surprising, because everyone's in the street. They are waiting for you. As you pass through the gate they begin to cheer, throw leaves and flowers. What have you done to deserve this? Ahead of you in the square is a group of men and women who await your coming. One of them steps forward and bids you welcome: 'We have waited for you. For so long there has been no king or queen in this city, but we always knew one would be sent to us . . .'. What have you done to deserve this, the crown of the city? But this, you are told, is Malkuth, your city. It can be no other. Yet the crown has to be earned, and to do this you must seek the High King. They give you a pair of sandals and a staff and they take you to the far end of the city and they show you the gate through which you must go. It's straight ahead, they tell you. And you begin your journey . . .

A straight, fine-built road, curves a bit, a few pot-holes – and suddenly the road isn't quite so smooth, and suddenly the sandals don't fit so well . . . Down into a dark wooded vale. It takes you into the wood. So dense that you have to push your way through the undergrowth . . . [This is a form of initiation] . . . you emerge the other side.

Before you are hills . . . Climb to the highest point of the hill. See before you a lake. It looks very inviting. You think about bathing your feet . . . you step down, but the water steps back . . . You try again but the water goes away. You think about this. Why? What is water? Water is a part of you. You are 90 per cent water. How about the water in you calling to the water in the lake? When you step in the water this time it's cool and stays where it is. You walk forward – and suddenly you're in over your head. You can breathe easily – let the water in your body become the water in the lake and allow yourself to sink down and you will find there a castle – the castle of the Lady of the Lake. She greets you at the door with a great silver cup. In the cup is milky fluid, the blood of the moon. The stuff of which dreams are made. Take it and drink it down. [This is **Yesod.**] Thank the Lady and let yourself rise up and emerge the other side of the lake.

The path has changed, it is rainbow-coloured. You have left your sandals – go barefoot. You have passed into a different level. The sky is a deep indigo shot through with different colours. You come to a crossroad, and there are two figures awaiting you. [This is **Hod** and **Netzach.**]On your left, a young man with an orange cloak wrapped closely about him. His eyes are steady as they look at you. On the right a woman is wrapped in a green cloak. They ask if they may travel with you. It is your decision. You begin to step forwards, but a sound makes you turn. A dog stands behind. It is like a wolf, grey and shaggy. It sits with its head on one side . . . You set out with your companions on either side, the dog behind. You can hear the sound of air, rain, wind and storm upon you . . . One companion offers you a cloak which is it – orange or green? It makes a difference . . . Move on. You see a rainbow arching in the distance. You hear the tolling of a great bell.

The dark indigo sky gives way to golden sun. [This is **Tiphareth.**] It seems to rise out of a dip in the horizon and shines for you. The landscape is illuminated. There is a chapel open to the sky, ruined. Set around it are ancient grey stones. In between a white horse gently crops the grass. In the doorway there is a man, behind him, standing beside an ancient altar, is a woman. He is dressed in a white robe, she is rainbow-coloured. He is Columba. She is of more ancient times. She is Brede, a fire Goddess. Out of the fire comes a creature, a salamander. She lifts it and it nestles in her hands . . . Match the heat of your body to the fire of the salamander and take it into yourself to combine it with your own inner fires . . . let it become a phoenix within you. The way of harmony takes both ways – Within you the water of the moon and the fire of the salamander. You have passed into a second level.

Pass through the ruins and go on. Come to a second crossroads. Two more figures await – a knight in red armour with his visor closed, and a woman of mature years dressed in blue and carrying a pair of scales.[**Geburah** and **Chesed.**] They ask if they may accompany you. Yes or no? Behind, the dog still stands guard. Begin the long straight, narrow and thorny road – gravel, pebbles and stones cut your feet. To the sides are small narrow pathways which appear smooth. If you wish you can take one. But you will do better to walk this one that lies ahead.

. . . the road is suddenly open and clear. But it is difficult to breathe, like high on a mountain. The road is upward, steeper, winding, going between high passes. There is cold biting snow. Your companions are silent but follow. Before you there is an abyss. [This is **Däath.**] Look down. You cannot see an end. Look across. It is too far to jump. You turn to look back at your companions, but they are not there. They are on the other side of the abyss. How are you going to get across? But get across you must. The dog is still behind you. It jumps up, and could push you over. Sometimes being foolish is a good thing. Being pushed over makes you see things in a different light. Light can become a bridge if you let it. A realization, a way between two points. Strange how you didn't notice that gossamer bridge before – like glass. The wind buffets you from both sides. The elements of air. Let it seep into you . . . Let it flow through you. The air emerges through your lungs like two great wings that lift you over the rest of the abyss. Earth, water, fire and air – you are all these things. From Malkuth use your wings to lift you over.

On the other side of the abyss two more figures – one wrapped in black, the other in grey [**Chokmah** and **Binah**]. Together you move on. Before you is a gate. Two great pillars with silver gates swings open. There come two figures, one is radiant [**Kether**], one is a man bearing a crown. Will you take the crown or will you refuse it? Will you rule or will you be ruled? Will you accept responsibility or will you lay it aside? Choose.

Move on through the gates. Your companions close around you, support you – Michael, Hanael, Raphael, Kamael, Tzadkiel, Tzaphkiel, Raziel . . . Angels of the presence escort you into the city of Kether, or is it Malkuth? For surely Malkuth and Kether are one and the same. The dog rears up to become Gabriel – the great indigo wings enclosing you.

Will you make this Kether your Malkuth and dwell here? Or will you return to humanity and tell them what you have seen? The decision is yours. You may rest here. You may take what is offered for yourself, or you may go back and tell others about the Royal Road. Make your choice. If you decide to go back then he who gave you the crown will give you a pair of sandals and a staff. Back to the abyss and this time of your own free will let yourself fall down into the darkness, back to earth, but not the same. Within you exist the elements of water, fire and air – given as gifts. Mix these with your humanity and the element of earth. Know this, that whether you refused it or not the crown is still yours – rule your own inner king and find your own divinity . . . it is within your own power to fulfil your destiny as a human being. Tread the royal road until one day you do not need to return but go on to the light. Back to the warm darkness, drifting and dreaming. Heavier. Begin to feel your body about you . . . Deep breath, and return.

When Dolores had finished speaking there was a long silence as people readjusted to ordinary consciousness. I had started a conversation with an ex-wiccan at the beginning of the conference, and we sat together for the pathworking. He turned to me and said how powerful the experience had been. There was a strong evangelical feel to this, and the message was clear: we had a mission to use the gift that had been revealed to us. It was our responsibility and duty to take the Royal Road to work on humanity's evolutionary consciousness.

The Royal Road pathworking above focused on the central sephiroth of the Tree of Life – from Malkuth to Kether. The Middle Pillar is seen to be the introductory or preparatory means of 'aligning the personality with the inner self, of identifying and unifying all the levels of the true consciousness which we, in our complacency and blindness, choose to call the Unconscious' (Regardie 1991a [1938]:132). The apprentice magician learns to build the Middle Pillar as meditation exercises – focusing the mind on the sephiroth of Kether, Däath, Tiphareth, Yesod and Malkuth, and building them up in colour in the mind's eye like a column. Different breathing exercises are practised, called 'The Circulation of Force'. The Middle Pillar is located between the Pillars of Mercy and Severity on the Tree of Life. These two pillars are symbolic of all duality; the magician must find the point of balance between them. When meditation on the sephiroth of the Middle Pillar is mastered the magician vibrates the God names associated with each sephirah. This 'Vibratory Formula' technique is said to awaken a power, or level of consciousness, to contact the corresponding force in the external world.

Spiritual Evolution: Quality Attracts Quality

> Never before have we faced such a crisis as that which now confronts us. As a Lifewave we have made too many mistakes, most of them due to our inability to live and work together in harmony. We are destroying the very ground beneath our feet. You may ask 'What can I do, I am only one person' . . .

These words were printed on a handout for a high magic 'Candlelit Vigil for Gaia' held at the annual Conference 1992. They point to a crisis due to lack of harmony but they also give the following solution: 'The answer is . . . "You can do a lot, if three such people get together they are a small group, if three such small groups get together they become a larger group. If six large groups get together they have a voice. When the voices are all saying the same thing they can become a power for good". . .'. The message is that I, as a high magician, can change the world by getting together with other magicians to work for the power of good. First, however, I have to start on myself – by changing myself to find the Divine Spark within. In the words of one adept, I need to 'build my own temple within to come face to face with my own cosmic centre'. This involves forming my own personal inner contacts of angelic or archangelic forces who act as teachers on the astral plane. This adept, an ex-Jesuit priest and ex-mason, stated with authority that, 'It is ordained for humans to be born in a balanced world and live a happy balanced life. By changing yourself, you can't help but influence the world.' Thus I make the choice to improve myself – through magical training – and thus further spiritual evolution. In fact, I have a moral duty to do so.

The significance of the concept of inner spirituality in relation to reincarnation and spiritual evolution was demonstrated to me at a weekend high magic seminar organized by my occult school, led by the co-leader of the school, and held at the London temple. A very large terraced house in a once very respectable neighbour-hood, now looking as though it had seen better times, is the temple of a number of high magic lodges. About five magical groups, each under the direction of a Magus, used the Temple at different times. On occasions they would meet to work together. The house did not look very different from the others in the road, apart from being a little run-down; however, it still retained its air of grandeur. A charming middle-aged woman, whom I recognized from the magical conference, opened the door. She let me in and told me where to leave my coat, and then led me into the back room, which was the temple. All the participants were gathered. It took me some time to adjust to the flickering candlelight and take stock of my surroundings. It was like stepping into a time-warp – I felt that I was transported into the 1920s and a very different age; and that I had entered into a very secret world. I listened to the outside noises in the street and contrasted them to the inside esoteric meeting. Seventeen magicians sitting around the altar with its central light and surrounded by candles and Egyptian artefacts had come to a workshop on the preparation of the 'temple within' – the 'sanctum sanctorum'.

The participants were largely middle-class, and most were experienced and high-rank high magicians, including supervisors. There was a hypno-therapist from Southampton in his late twenties, who was concerned about people in his lodge playing with what he saw as 'dark forces', and an older woman in her late sixties. A number of participants worked in the area of health: one was a homeopath, another an osteopath and another, an ex-mason in his late thirties, worked in a London hospital. A Portuguese woman also in her late thirties was experienced in tarot and colour therapy, and had just started the correspondence course, but was finding it hard to become sufficiently disciplined to meditate every day as required. A witch of twenty years from Manchester said that she had started on the preliminary lessons, but was not sure if it was for her. She had been persuaded to try it by her dowsing husband, who was also there. Another younger man was also interested in dowsing and researching Dion Fortune. He was very sensitive to the energies of places. There was a Nigerian man in his mid-twenties whose father was an Anglican minister, but apparently had practised magic secretly with the Rosicrucians.

The adept told us to be truthful to ourselves, to meet ourselves, to face-to-face to find our inner cosmic centre – our sanctum sanctorum. The idea of developing a sacred cosmic centre was to discover personal contact with the hierarchy of forces. This would lead to changes that would directly affect ourselves and the world – 'By changing yourself, you can't help but influence the world' – a magician's attitude will project on to the immediate environment and be beneficial to all. Everything was interpreted in terms of karmic balance – self-improvement through re-incarnation – 'we awaken on this plane as the totality of our experiences' – to work out our inner potential. Blavatsky's notion of spiritual evolution was emphasized. We were told that 'dogma restricts', and we had to find our own light and make our own laws; but these laws had to conform to Divine Laws.

We were encouraged to create our own temple within – to be our own priest or priestess within – to meet the self face-to-face at the cosmic centre. It was explained that there were four archangels in the internal temple: in the east was Raphael, represented by the Hermit, who was a guide to illumination of the path; in the south Michael, represented by a lion, embodied qualities of strength and courage to face adversity; the west concerned how the individual related to his or her anima or animus, was symbolized by the High Priestess (for men) and the High Priest (for women), and was the domain of Gabriel; finally, the north was expressed by the tarot card of The World and concerned everything in balance and unison. This quarter was the realm of the archangel Uriel.

The adept led a guided meditation to help us find our contacts. Afterwards we were told to make the four archangels part of our family, and to personalize our relationships with them as four angelic forces that would become our teachers if we let them make their own meaning. A discussion ensued about the symbolism of the four forces. A comparison with witchcraft was made, and one woman said that she could not work with four masculine energies. The adept told her that by making her own association with the forces she could personalize them in any way she wished, and that the qualities he had talked about were his own interpretation.

'The world has gone mad', the adept said. 'There is famine and disease. Is there a logic?', he asked. He then proceeded to tell us that by discovering our sacred centre – our inner sanctum – and invoking the angel Raphael we could bring healing to counter 'negative vibrations', which are apparent in everything from bad thinking to polluted food. Everything was prone to negative vibrations; but if the physical vehicle was brought to a state of sanctity within the light it could re-charge bodily particles to bring about balance, peace and order.

Shortly afterwards he led a pathworking connecting us to our inner sanctum. We were then each asked what we had experienced. I spoke about how, even as a young child, I have always wanted the world to be peaceful, and my experience of this pathworking was to connect my peaceful microcosm into the troubled macrocosm. He asked me if I would consider joining seven ordained magicians who were working on world peace. This came as quite a surprise to me, and later led me to consider my position within the occult school. The invitation to join a special sub-group of the school dramatically increased my status within the group of magicians. Previously, I had been ignored by some of the older participants who were supervisors; but after the offer made by the adept they started talking to me and asking whether or not I would accept the honour.

This change in their manner clearly demonstrated the hierarchical system on which such magicial knowledge is based (see Chapter 5). As a beginner I was not considered worthy of association, but as a potential initiate of an inner order, I suddenly became worth knowing, and magicians previously distant came to talk to me at the tea break afterwards.

Seeing the Wider Pattern

How does high magic affect the magician's view of the socio-political world? In theory, the Tree of Life represents both humanity and the cosmos, and is projected on to the socio-political world. Thus the socio-political world is interpreted and explained by this glyph, and the magician comes to believe that magical practice on the microcosm (the self) will affect the macrocosm – which includes the ordinary everyday world that the magician inhabits. The Tree of Life becomes a reference point, a screen through which the world is viewed. Thus contemporary high magical practice is concerned with developing the magician's consciousness to correspond with the consciousness of divinity. It is through an understanding of its symbolism that humanity can discover its divine nature. The Tree of Life is described by Fortune as a method of using the mind to aid humanity's spiritual evolution; it is a composite symbol representing the cosmos and the soul. Each sephirah represents a phase of evolution. The paths connecting the sephiroth represent subjective consciousness, by which the soul realizes the cosmos, thereby raising evolutionary consciousness (1987b:17).

In theory these are high ideals based firmly on a notion of evolution. I wanted to know how these ideas worked out with ordinary Pagans in the subculture. I had got to know a group of Pagans from all magical persuasions at a monthly social/magical talk evening. We met in a Pagan couple's flat in south London. The small room was bulging to overflowing with the local Pagan crowd, and the atmosphere was very friendly. The evenings were usually well-attended, and were lively and relaxed. Each member would take a turn to give a talk on a subject that interested them. This was a chance for members to do some research and reading and to get some comments and discussion. There were talks on Chaos Magick, Divine Kingship and Aromatherapy. I contributed a talk on the 'Underworld' one evening, and spoke about how all descents to the underworld provided entry into different levels of consciousness. There was a lively discussion afterwards, and we spoke about how the language of the underworld was non-verbal, and shaped by myths and symbols.

On another such occasion a talk on the Kabbalah was given by Matthew, an experienced magician from my magical school. He had spent a number of years working magic, and had a great deal of experience in magical ritual. He was taking time out from his contact with the occult school to 'follow the path of the mystic'. He explained that the mystic treads new and untrodden paths, while the magician treads well-trodden routes through ritual practice. He is trying to integrate all world religions, plus witchcraft and chaos magick, with the Kabbalah.

The usual gathering were there – our Pagan hosts, Richard who was a Pagan chaos magickian and his partner Cathy, both in their mid-twenties and with four young children (Cathy was a 'hereditary' witch, and ran her own coven and training class, of which Claire and Martin were members); Peter, a solo witch;[5] Frank, who was an older witch from a Gardnerian coven; and two new members, Kate and Helen, who were interested in magic. Matthew had brought a profusion of books on the Kabbalah, and they were laid out on a small coffee table. People picked up the books and flicked through the pages, discussing their relative merits and asking Matthew questions. The table was later cleared for crisps and cartons of fruit juice.

Matthew told us that everything in high magic was about balance. He showed us some exercises that are done before or after ritual for balancing, calming and attuning to 'the forces'. The first was the Isian meditation, which we were invited to try. I lay down on the floor between Martin, who knelt with my head in his lap, and Matthew, who held my feet. Matthew's energy was circulated from Martin's Yesod (the second sephirah on the Kabbalistic glyph) through my head down my 'middle pillar', as I lay on the floor, and out to Matthew's Yesod and up and across to Martin and down to me again in a circular motion. Then Frank and Helen tried a different exercise. Frank knelt before Helen and placed his forehead on her Yesod and circulated his force from Helen's Yesod down through her and up his 'middle pillar' through to her Yesod again. After a lot of experimenting and discussion we settled down to a pathworking by Matthew, where he took us on a meditation route similar to Dolores Ashcroft Nowicki's Royal Road already described.

When we had all returned to ordinary consciousness, Matthew explained how the concept of evolution was absolutely central to high magic practice. He said that all life forms are evolving – animals, plants – everything was on the Tree of Life, whether conscious of it or not. This provoked an interesting discussion, because I had been to a seminar run by my occult school that took Dion Fortune's view that human beings were the supreme carriers of divinity, and consequently had a responsibility to elementals.[6] This was because elementals were not created God but by humans – who do not have the 'power to endow immortal life' (1987) – and the elementals were therefore soulless. The only way they could become ensouled was to work for it. Thus human beings were seen to be the 'redeemers of matter' and could 'adopt' elements to work with. This helped the elementals, and, in return, it was said, the elementals would perform services for the magician. In Fortune's view humans are the supreme carriers of divinity. We all have the potential, but only some of us are called to service. Following the high magician's path is seen to be an elected process for those who 'choose to take the function to improve themselves or to take the knowledge to improve a group of people'. Matthew disagreed with this. Everything evolves at its own rate, and it was not a question of 'us helping them to achieve souls', he said.

This raises interesting issues about evolution and hierarchy; and magicians appear to have conflicting and often confusing views on the subject. It is my impression that much of the original work of Dion Fortune – which was itself a development of Blavatsky's notions of evolution – is now viewed as openly élitist, and is being re-interpreted in a more democratic fashion. In addition, there seems to be a general confusion between Darwinian notions of physical evolution – a progression from simple organism to human – and the notion of spiritual evolution advocated by Blavatsky.

Nevertheless, theories of evolutionary consciousness determine how the high magicians come to view the world. The source of divinity is part of the world (Kether is in Malkuth and Malkuth is in Kether: 'As Above – So Below' is an oft-quoted maxim). However, the concept of a 'Fall' from divinity creates a gap, a distance between creator and created, which is bridged by notions of evolution, karmic cycles and reincarnation, all of which demonstrate the abiding influence of Blavatsky and the Theosophical Society's teachings. These ideas raise contrasts with the witchcraft view of the world, where the earth and humanity are inherently sacred. These positions have interesting philosophical roots (Löwith 1967), which will also be discussed in greater detail in Chapter 7. In witchcraft the world is interconnected energy that is both harmonious and divine. By contrast, the task for high magicians is to evolve consciousness and bring divinity down to the world.

These evolutionary theories do not deal with history or issues of power – absolutely everything is seen in terms of the Tree of Life. Politics are seen in terms of imbalance. One magician told me that, in the case of famines, we must not create more imbalance by throwing food at people, as it 'will only encourage them to breed more and create more imbalance without any resources'. It would be better to do magic to make the crops grow. Thus the politics of the ordinary world are irrelevant to the Tree of Life. The Tree is seen to be central – everything works out in its own course like a biosphere – 'Gaia does not tamper with the rhythm.'

For high magicians in this tradition, then, mastery of oneself is central to magical work concerned with expanding what is perceived to be a higher consciousness of divinity – a human reunion with divinity. This involves what is seen as an emergence from nature and baser instincts and the lower self to a harmony with cosmic law.

The End of my Apprenticeship

The politics of this magical approach made me reflect on the basic teachings of the occult school. The experience of being invited to work on world peace made me consider my magical training, the anthropological fieldwork and how much I was willing to undergo in the name of the research. I wrote to my academic supervisor: 'I feel really confused by how I feel about some of this stuff as in the very broadest sense I do resonate with *some* of their principles and doing path-workings' (Report from the field no 17, 1 February, 1993). After much reflection on my feelings about this magical school, and also on my position as an anthropologist, I wrote to the adept explaining that I felt very honoured to be invited to research the causes of the four plagues on the world, but that I was not sure whether the school was the right path for me. I also told him that I was conducting anthropological fieldwork exploring different magical spiritualities arising from a long-standing interest in magic. I did not receive an answer to my letter.

Consequently, although I had been invited, I decided not to take initiation into further work with the school. I was experiencing difficulties with the practice – it was becoming increasingly hard to work each day on the meditations – and I had decided that this was as far as I was willing to take my apprenticeship. I decided that it was time to write to Pearl about how I was feeling about the training:

While I am aware that deity is viewed as female as well as male and that it goes way beyond gender, I am not happy working with some of the lesson material which I think is very patriarchal – and to be honest it just doesn't work for me. I know that the earlier lessons are about developing a strong personality, so that when proper magical work is undertaken, the character is strong enough to hold and mediate the force. I understand this, and realise that application, dedication and hard work are the vital elements on the magical path. I am willing to work, but cannnot resonate with much of the course material.

Inevitably, Pearl's response was somewhat defensive:

One thing that has to be born [*sic*] in mind is that the course was written a long time ago and the reason that it has survived for so long is that it is SO good and produces excellent results when it is followed to the letter of the word. If, however, you find yourself choking on the word, then it probably isn't the right course for you. As I said when I first wrote to you, there are many paths but only one destination.

She suggested joining a wiccan group, whose emphasis on the Goddess might be more suitable for me. I was left feeling that there was little room for personal exploration outside the rather rigid structure of the school. Pearl's comment that the course provided excellent results when 'followed to the letter of the word' seemed to sum up the general fundamentalist approach. It was difficult to know if this was a reflection of her as a teacher or if it was the ethos of the school. Certainly I had had conversations with other supervisors, and they seemed more open-minded. Equally, I had met others who were probably even more pedantic than Pearl.

My reflections on the training are that the overall Kabbalistic structure of the Tree of Life glyph provided a useful framework for experiencing various subjective otherworldly states. However, that framework had been subject to a particular interpretation – one that I personally found difficult to work with. Moreover, the lack of engagement with ordinary reality fostered a spiritual escapism I found hard to handle.

Thus my training in a high magic school terminated. In the next chapter I examine the witchcraft approach to magic and the otherworld.

Notes

1. Because of her view that magic is a pseudo-science, Luhrmann is preoccupied with the issue of scepticism and seeing how magicians come to find magic 'sensible and realistic'. She points to the way that dynamic experiences become

part of the business of engaging in magic, and maintains that this is what makes magic real for its participants, by giving content to magical ideas. She claims that magicians protect their involvement by a range of *ad hoc* 'patch-up job' arguments that allow commitment without violating their scepticism (1989:12).

2. High magicians organize themselves into groups called Orders, Lodges or esoteric schools for the purpose of working with the forces of the otherworld.

3. The application form consisted of 10 questions. They ranged from whether I belonged to any occult or esoteric group or society, what my motive was in studying the Kabbalah, what occult books I had read, and whether I had attempted or carried out any occult, Kabbalistic or ceremonial ritual or magic, to whether I had any skills or could speak any other languages. It also asked whether I would be prepared to accept the Principal's ruling on the administration of the Order as final, and if I would be prepared to carry out work for the Order if requested.

4. The origins of the Kabbalah are contested, but most magicians believe it to be a Hebraic mystical system.

5. Solo witches are sometimes called 'hedgewitches', and often practise herbalism and folk traditions associated with the countryside.

6. For Dion Fortune elementals have no Divine Spark and will be disintegrated at the end of this evolution, unless they can develop a spiritual nature. They are a result of a series of 'constantly co-ordinating actions and reactions' or 'tracks in space', which remain after the activities that gave rise to them have ceased (1987a:47).

—4—

Witchcraft and Natural Magic

Witchcraft rituals tend to be conducted outside in woods or open spaces; the emphasis is on a connection with the land and its spirits rather than the angelic beings favoured by high magicians. In the preceding chapter I was concerned with the experience of training as a high magician and an explanation of notions of spiritual evolution and the bringing down of higher angelic forces into the magician. This chapter will also focus on my participant observation, this time within three witchcraft covens, and will address the issue of 'natural magic'. Paganism is said to be a nature religion, and witchcraft is seen to be the magical practice most closely associated with nature. Drawing on Margaret Murray's theory that the witch cult of Diana was a pre-Christian fertility religion practised chiefly by 'the more ignorant or those in less thickly inhabited parts of the country' (1921:12), Gerald Gardner has created a modern version of witchcraft that combines elements of high magic and folk practices.

The chapter starts with my ethnographic encounter with witchcraft through a description of various rituals, and is followed by an examination of invocation – the drawing down of particular forces of the otherworld, usually as a goddess, into the high priestess or other female witch. This leads into a discussion of the importance of women in the practice. An analysis of witchcraft as nature religion completes this section. I question some of the implicit assumptions of the practice, by asking how much practitioners are actually engaged in a relationship with nature and how much is lip-service. For all magicians the otherworld is internal as well as external in the wider being; however, notions of nature are largely shaped by the Hermetic tradition of the Renaissance, which emphasized human nature – that is internal nature – rather than external nature. Modern witchcraft is not so much an indigenous nature religion, but rather a less formal development of high magic focused on the veneration of femininity. The chapter concludes with a comparison of witchcraft and high magic.

An Introduction to Wicca

My opportunity to speak to some wiccans came when I got talking to a woman at a high magic seminar whom I shall call Sarah. She told me that she was a wiccan high priestess, and I asked her if I could join in her coven's rituals. After the seminar I went back to her flat to meet her high priest, who was called Phil, and another priest so that they could decide if I was suitable to join the coven. We talked about the relative merits of high magic and witchcraft, and discussed books. As with most magicians, their flat was full to overflowing with books, and Phil said that it had reached the stage where the books (mostly on magic and related subjects) were stacked on the shelves in double rows, one row in front of the other, owing to lack of space. We talked about magic and my views on life in general. Some time later they told me that the next ritual was for Imbolc (Candlemas), held on the second day of February, and they invited me to join them.

This wiccan coven was fairly small: it was based on an established high priestess and high priest relationship and three or four regular members, all of whom were highly educated and had professional jobs in research or in health care. Sarah, the High Priestess, was in her early thirties and worked in the medical profession. She was an intense woman with a keen intellect, and was in the habit of asking deep questions about the nature of magic. Phil also had a quick mind and a good sense of humour. He too worked in health care. I got the impression that he came from a reasonably well-off family background, because he said he had enjoyed playing rugby at school and competing in yacht races from the Isle of Wight when he was younger. He explained how he had been inspired by being with the elements when he was sailing, and how, on such occasions, you have to get it right, because 'the balance between life and death is very fine'. Phil and Sarah had got into witchcraft through meeting someone who had been trained by a hereditary witch in Scotland. When I asked what hereditary witchcraft was Phil answered that it was like wicca, but much more secretive. Phil and Sarah had since developed their own brand of witchcraft by working on their own magical ideas.

The coven had what its members termed a 'special grove'. This was a small clearing in a wood a few miles from their home. This site had its own energies, and if the spirits of the place were unwelcoming then the ritual was abandoned.

The Witchcraft Circle

The witchcraft circle is a legacy from high magic, and in Britain it is usually arranged in the following manner: the east represents light – the first light at dawn – and also spring. The circle is usually opened in the east, which is symbolic of air, intellect and rational thought, which classifies and divides. It is symbolized by the athame or a sword. The south represents fire – the energy of the magical will. It is symbolized by the wand, which is a slender branch of wood – hazel being especially appropriate, since it has the capacity to bend. The south is representative of summer, sun and heat. This is the quarter for energizing, for the realization of the will – a course of action. The west represents water, the emotions, rivers, oceans, streams and autumn. It is symbolized by the cup or chalice. The west flows and merges, while the east separates. The north represents the earth, mountains, valleys, winter and the body. The pentacle is symbolic of the north, the five points representing the five senses of the human body, and all the four elements – earth, air, fire, water – plus spirit. The north corresponds to darkness and winter and internal reflection. In the centre of the circle stands a large cauldron, which represents the change required to transform the raw into the cooked. Doreen Valiente, Gardner's High Priestess, explains the symbolism of the cauldron:

> A cauldron is an all-embracing symbol of Nature, The Great Mother . . . the vessel of transformation, because it takes raw uneatable things and transforms them into good food; makes herbs and roots into medicines and potent drugs; and is the emblem of woman as the greatest vessel of transformation, who takes the seed of man and transforms it into a child. In a sense, to the pagans all Nature was a cauldron of regeneration, in which all things, men, beasts, plants, the stars of heaven, the lands and water themselves, seethed and were transformed (1986:58).

The focus for contemporary witches who celebrate rituals based on Gardner's structure is to link the body, and bodily experience, with the spirit world of otherworldly reality. To stand in such a circle is to be very aware of the all-encompassing totality of existence. The circle contains and represents the whole, while the cauldron and its central light represent eternal spirit, which entices the practitioner to look within. The emphasis is on connecting the internal with the external. The witchcraft circle is the place where physical, social and spatial boundaries are redrawn. The circle represents the wholeness of the human, the natural and the divine. All are seen to be incorporated into one, in a reunification of matter and spirit. The breach, which according to witches was created by Judaeo-Christian traditions, is restored. Magic provides the healing threads. Participants, by entering the circle, become the centre of the cosmos. The circle is a 'meeting place of love, joy and truth', above all a place to find the lost fragmented self. Nature represents the antithesis of urban society and its Christian values. Witches create what they see as a holistic pre-Christian world, and employ magical techniques of altering consciousness in an attempt to heal the rifts of a dualistic culture. The general magical view of the body inherited from the Hermetic tradition is that it is a locus of the macrocosm. This means that the individual body contains all the energies and forces of the cosmos. In essence it is a field of energy that has the potential to channel the forces of the entirety of the macrocosm. The Goddess is seen to give birth to the world, and so deity is present in the world – in matter and the human body.

Imbolc Ritual

I joined Sarah and Phil's coven for the Imbolc ritual at the special wood. It was seven o'clock in the evening, and coveners had started arriving at Sarah and Phil's flat to celebrate the Imbolc ritual. Sarah was ordering everyone around – the wooden bowl for the apples needed oiling, the drums needed dampening so that they would not split. People took their turns in the shower – it is obligatory to wash carefully before any ritual work. They emerged wearing towels and made their way across the crowded floor to change in the adjoining room, reappearing fully robed and transformed into ritual specialists. The ritual paraphernalia – cauldron, lanterns, salt and water, compass, and incense – was checked and placed carefully into cardboard boxes before being taken to the cars. We sat down and discussed the plan of action. Everyone was made aware of their special part in the ritual drama. Then we piled into two cars and made our way to the coven's 'special grove'. Tension and excitement were high. On arrival the cars were parked and we started the long walk to the grove. Everyone was wearing long flowing robes and Phil was carrying a long staff – we must have made an interesting sight to any casual observers as we walked to the wood. As we got further from the road, so the ordinary world receded. We were entering a magical world. It got quieter and darker. We slowly made our way down a long avenue of trees and I was introduced to 'the guardian', a fallen bough, which loomed out of the dark like an elongated dragon. The darkness took on a life of its own, of shadows, strange damp smells and whispering noises, as the ordinary world seemed to get farther and farther away. Some went ahead to set up the circle; others took their time absorbing the atmosphere and energies of the place. We eventually reached a little wooded copse of mainly young oaks, which was set down in a dell, a natural clearing. Crows were cawing in the trees, and there were rustling sounds in the undergrowth. The moon was waxing three-quarters full, and a light misty glow made the wood seem very magical. The place had a peaceful feel to it.

I was introduced to the trees and the four quarters. We spread daffodils around the circle deosil (clockwise). Then came the purification by Sarah the High Priestess:

> I exorcize thee, O creature of water, that thou cast out from thee all the impurities and uncleanliness of the spirits of the world of phantasm; in the names of Cerridwen and Herne.

The purification for the salt by High Priest (Phil):

> Blessings be upon this creature of salt; let all malignity and hindrance be cast forth hencefrom, and let all good enter herein; wherefore do I bless thee, that thou mayest aid me, in the names of Cerridwen and Herne.

Salt-water pentagrams were painted on all the participants' foreheads. A sage incense 'smudge' stick was lit, and the smoke was gently traced around our bodies. The salt and water were sprinkled around the circle to purify it.

The High Priestess opened the circle with her athame (a small ritual knife):

I conjure thee, O Circle of Power, that thou beest a meeting place of love and joy and truth; a shield against all wickedness and evil; a boundary between the world of men and the realms of the Mighty Ones; a rampart and protection that shall preserve and contain the power that we shall raise within thee. Wherefore do I bless thee and consecrate thee, in the names of Cerridwen and Herne.

The watchtowers of the four quarters were then built in turn. The coven faced each direction, starting in the east (the place of sunrise) and drew an invoking pentagram with outstretched arms:

Ye Lords of the Watchtowers of the East, ye Lords of Air; I do summon, stir and call you up, to witness our rites and to guard the Circle.

Ye Lords of the Watchtowers of the South, ye Lords of Fire . . . [and so on]

Ye Lords of the Watchtowers of the West, ye Lords of Water, ye Lords of Death and Initiation . . . [and so on]

Ye Lords of the Watchtowers of the North, ye Lords of Earth; Boreas, thou guardian of the Northern portals, thou powerful God, thou gentle Goddess; we do summon, stir and call you up, to witness our rites and to guard the Circle.

Three of the women, representing the Maiden, Mother and Crone aspects of the Goddess, positioned themselves in the centre of the circle around the cauldron, which contained a lantern. Each wore a veil of the appropriate colour – Maiden, cream; Mother, red; and Crone, black. Invocations were chanted to Cerridwen and Herne (the coven was working with Celtic goddesses and gods at this point) accompanied by drumming. The other coveners were positioned around the circle. We had each chosen a 'special' place. I positioned myself between two low branches of a small tree in the south-west (afterwards I learnt that it was a very powerful tree that was 'between the worlds'). Some of the other participants became boulders or stones. The Maiden came with the lit lantern and awakened us individually to the first day of spring and a celebration of the light that was lit within each one of us. We started dancing and drumming and shaking our rattles around the Triple Goddess in the centre as power was raised. Then we held hands in a circle, male, female, male, female, and sent the energy to the Shetlands to help the environment cope with an oil tanker spillage disaster. The Maiden, who had sensed when the power had reached its peak, called a halt, and we channelled it to help the ecological crisis. Then apples and ale were shared amidst joyous hugs and celebrations. One apple and some ale were left for the earth. The watchtowers were thanked and banishing pentangles drawn in the air. The coven slowly retraced its steps along the pathway, amidst wolf howling from the High Priest, back to the ordinary world. The waxing moon illuminated the way.

I was greatly relieved that the ritual had gone well. On the day of this ritual I had been suddenly seized with panic, and in my fieldwork diary I had asked myself why I was taking myself off in the night to the middle of some wood with a strange group of people who could have been sympathizers of Jack the Ripper. My fieldwork report to my academic supervisor records that I had considered crying off. However, I decided that it was an opportunity that I did not want to miss, so I tried some of the magical techniques that I had been taught: 'I asked Sekmet, an Egyptian lioness goddess, who is one of my four quarter guardians, for strength – and it certainly worked! I went to the ritual and I was nervous, like before any ritual, but I had contained my paralysing fear . . .' (Report from the field no 18, 1 March 1993). In her response to my report, my academic supervisor asked me why I had felt such fear about doing the ritual. This prompted me to think more about it, and in my next report I wrote:

> Primarily I was scared because I did not know the group and I felt a little uneasy about them. Also interference and possible hostile reaction is a problem with working outside (after the ritual in the wood there was a distinct feeling of relief that all had gone well). I spoke to one of the men participants and he said that he always felt nervous about possible interruption. He said that they didn't work too often in 'their sacred grove' because if the place got known there would be trouble (Report from the field no 19, 1 April 1993).

A confrontation with such fears and an opening up of oneself to the fieldwork situation was part of the process of this research. It meant putting myself in potentially unsafe situations, which I probably would not have done in ordinary circumstances.

Into the Coven

After the Imbolc ritual I was invited to dinner with Sarah and Phil and another coven member named Ben. It was a time for them to find out more about my views on magic. I arrived at their flat in the early afternoon and was welcomed inside amidst a flurry of tidying-up activity. Ben had just come in with some shopping and he and Phil started putting it away, while Sarah started cooking the meal. She asked me what I had been doing since I last saw her, and we started talking. Ben started peeling and chopping vegetables. During this time Phil straightened a bent knife with a hammer. He had carefully unwrapped the hammer from a piece of blue silk whilst talking about the sacred art of the smith. When he had finished, he placed the hammer back on the piece of silk and wrapped it up.

After the meal Sarah asked me for my opinion on the meaning of ritual, and which esoteric text I found the most useful. During the conversation the subject of psychotherapy was raised, and Sarah said that psychotherapy was vitally important because an unbalanced person could not mediate the forces. However, she said that there was more to magic than this – magic was about mediating the forces in a ritual circle. They said that they had a 'bun fight' meeting before any ritual to sort out personal differences, and discuss how they were going to do the ritual. Phil spoke about how a person entered a circle in 'perfect love and perfect trust' and how the group had a responsibility to other members not to divulge practices at the workplace, or where they would be misunderstood. Phil said that anything could be brought into the circle and the group would listen and not judge. If the group had an unfavourable view about a person, they would speak to the individual, and it might be decided collectively that the group was not suitable for that person.

Sarah asked me if I considered myself an initiate. When I said that I was not initiated, she spoke about how she thought it was a useful way of describing oneself as apart from the more orthodox practices and wider society. Phil asked me if I practised any form of divination, and on my response that I was learning a little through tarot, they got out their vast collection of tarot cards, unwrapping each set from a piece of coloured silk. There followed a discussion of the relative merits of the many different types. Sarah and Phil said that they welcomed discussion and dissension so that they could learn, because theirs was a learning and growing coven. They said that they were looking for new members (although not too many, as the coven had always been small). They wanted commitment from coven members, and not people who leeched their energy.[1] Then we spoke about the details of the next ritual, the Vernal Equinox (sometimes called the Spring Equinox), which was going to be a celebration of the balance of the light and dark of the year.

In contrast to high magic rituals, which are almost always indoors (in a 'temple', showing the importance of the Masonic emphasis on building and architecture), most of the witchcraft rituals I attended were conducted outdoors in 'nature'. A close contact with nature – with trees and perhaps a river – was seen to be important, and could be made even in mundane circumstances such as in the middle of a park, crowded with people on a busy Sunday afternoon. This was an experience I had with Sarah and Phil's coven at the Vernal Equinox. The ritual developed in an interesting way, because a female member was ill with a migraine, and this upset the organization of the ritual. The group went through the alternatives. Someone suggested that we do some craft working together – carving, sewing or such like – but the consensus was that the majority would prefer to do another ritual of some description. So Sarah sat down and wrote another ritual, which involved making four small squares of wood – two dark and two light. She ordered Phil to set to, and without a minute to lose the workmate was installed in the living room and he was hard at work cutting two one-inch cubes from a silver birch branch and two cubes apparently from a 3,000 year old ('carbon-dated') piece of bog oak from Ireland. Eventually, about four hours later, Phil finished the four cubes of wood. As the ritual had been delayed we were unable to go to the original ritual grove, owing to lack of time. It was decided to use a closer location – a park. An 'undercover' working was decided on, so as not to attract attention.

When we eventually arrived at the park the ritual was performed in ordinary clothes and mostly in silence. Sarah as High Priestess chose a spot between three trees, by a river and next to a path where there was a semi-continuous stream of people walking. We were purified by her, using water from the river. We silently invoked the four quarters. I had a black cube and the High Priest was my counterpart with the other black cube. The two others, the High Priestess and the second priest, each had a white cube. We meditated on the significance of our cubes. Each had prepared an invocation the previous week – mine was inspired by the Gnostic Gospels and the commentary by Elaine Pagels (1990[1979]), and concerned the movement between light and dark, inside and outside, male and female. We quietly read our invocations. Then my dark cube was bonded with the light cube from my male partner. The High Priestess glued them together and we ceremoniously bound them with blue cord. The same was done with the other two cubes and then the four were joined together and bound. At the Autumn Equinox it was planned to carve a pentagram on them. As the ritual was drawing to a close a dog came bounding through the circle with a large branch, which it wanted throwing. Its owner looked embarrassed and not quite sure how to approach these rather strange-looking people. The stick was dispatched for the dog by one of the magicians, the quarters were silently thanked and a goblet of wine and some cress were shared.

Another Coven

At the same time as working in Sarah and Phil's coven I joined an evening class on Myth, Witchcraft and Magic run by Ken Rees. Ken started Ph.D. research on Paganism in the 1970s, but did not complete it, and now lectures on magic, witchcraft, shamanism and mythology. A first-degree Alexandrian and second-degree Gardnerian witch he has also been initiated into the *Ordo Templi Orientis* caliphate (an organization specializing in sexual magic), and has has worked with the well-known witches Janet and Stewart Farrar and Dolores Ashcroft-Nowicki among others. He introduced me to a wiccan coven with which he was involved. It was run on egalitarian lines – everyone took a turn at writing and leading the rituals. He said that he liked hierarchy until everyone had an understanding; then the emphasis was on sharing responsibility, with no single person dominating.

The rituals of this group were enacted as sacred dramas, and I was invited to their Spring Equinox celebration.

I met Ken a couple of days before the planned ritual, and he dictated to me the sequence of the performance (which he had written) so that I would know when to speak my lines and what actions I should perform. Wiccan rituals tend to be like dramas, with everyone knowing their role and what to say, although there is usually some space allocated for spontaneity too. He explained that wicca is a framework on which a magician could creatively hang any ritual performance. I went home to learn my lines and rehearse in my head what I had to do. On the evening of the ritual I drove to Ken's flat. We put the cauldron in the boot of the car, and drove to another part of London to pick up Rachel from her work (she was a nurse). We then drove to a house in a very smart area of north London, where I met the other two members of the coven: the owner of the house, Peter, who was in his forties, and Vanessa, who was an art student in her late twenties. The whole house had a rough artisanal feeling to it. It was a house of someone in touch with the forces of nature. Peter pointed out drawings on the wall that he had done when he was about four years of age – they were of witches. He was a trained artist, and the walls were decorated with his own oil paintings. He said that he had been inspired by Robert Graves's *The White Goddess*. The house was full of books about country life and witchcraft.

The ritual was going to be held in the large studio lounge, which had an ochre circle of thirteen moons painted permanently on the floor. The circle decorations were attended to, the cauldron was put in the centre of the room and filled with water and decorated with spring leaves and flowers. To the north was a rough-hewn altar laden with bowls for salt and water, flowers, a bell, a feather, Goddess and God figurines. Candles marked the other directions.

When everyone had changed, Ken cast a 'Gardnerian circle' and created sacred space, and I purified it with salt and water. The ritual went as follows (I reproduce it here from the copy that Ken gave me):

Rachel stands in the West; Ken stands in the North with stang (a forked stick); Peter stands in the East; Rachel in South; Susan remains in centre.

Ken:
Today, we stand poised between the powers of Light and the powers of Darkness. From today the powers of Light will wax and the powers of Darkness will wane . . .

'Farewell, now powers of Darkness
farewell, dark battlements of the Moon
farewell, still woods of Night
farewell, dark serpent of Dreams
farewell, Dark Mother.'

Ken now moves to the East standing with stang extended drawing solar power from the East.

Rachel (speaking from the South):
Today we turn to your son born of the deep Cauldron of the sky who shall rule us till the wheel turns again and we come back once more to your realms to seek your blessing.

Susan (speaking from the centre, addresses the cauldron):
Behold the Cauldron of Cerridwen, womb of the Gods and of men, awaits now the seed of the Sun.

Ken moves deosil around the circle from the East, finishing at the centre. He and Susan place their hands round the top of the stang and plunge it upright into the cauldron.

Ken:
The Spear to the Cauldron

Susan:
The Lance to the Grail

Ken:
Spirit to Flesh

Susan:
Man to Woman

Susan and Ken kiss over the cauldron. Ken retires to the West. Vanessa moves to the East, takes up the wand from the altar and invokes:

We kindle the fire today
In the presence of the Mighty Ones,
Without malice, without jealousy, without envy.
Without fear of aught beneath the Sun
Save the high Gods.

Thee we invoke, O Light of Life!
Be thou a bright flame before us
Be thou a smooth path beneath us
Be thou a guiding star above us.

Kindle thou in our hearts within
A flame of love for our neighbours
To our foes, to our friends, to our kindred all,
To all people upon the broad Earth.

By the music of the hills, I invoke thee
By the signs of the old stones, I invoke thee
By the phallus and the pine cone do I invoke thee
and call upon thee
To descend into the body of thy servant and priest
That thy people may know thee.

Vanessa draws a pentagram, sunwise, upon Peter with the wand. She continues:

Welcome, O Lord of Light,
The Wheel has turned full circle
Born anew you return with life,
From the secret places of sleep.

We of the Old Ways
Who alone are Children
Of the Light and the Dark
Rejoice with joy.

Vanessa then gives Peter the 5-Fold Kiss. This is completed with a kiss and an embrace, both grasping the wand, which is then left with Peter.

Peter:
Welcome the Powers of Light
Welcome the warm breath of Spring
Welcome the springs and the rivers,
Welcome the leaves and flowers of the Earth.

From a woman of the night I was born
From her womb has sprung the burning seed of the Sun.
I bless the seed with the Light of Life
And return it to the ground
That the year shall come forth and the light shall return.

Rachel leads snake dance around the Cauldron, deosil, picking up Ken, Susan and finally Vanessa in the West. Meanwhile Peter mimes the planting of seed in the ground . . .

Rachel:
Dance ye about the Cauldron of Cerridwen the Goddess and be ye blessed with this water, consecrated by the seed of the Sun, arising in his strength in the Sign of Fire.

About 3 circuits, after which all return to their earliest previous quarter positions. At some point Vanessa, surreptitiously, should have hidden the daffodils (or similar).

Peter:
I seek now my bride,
For this is the season of my arising.
She is in some fair glade dwelling
And in the sky her ship is gliding.

Vanessa:
He who would hath the Lady in *her* season must pass a test or two,
O suitor, find for me the flowers of the spring
So that I may recognize you on eve of May.

Peter and Ken start a ritual search for the flowers. Whoever finds them presents them to Vanessa with a courtesy and a kiss.

Vanessa:
You have done well. Now, within 2 moons I shall recognize your valour.
Yet you will still have a test to come.
For then I will wear a different face
In the meantime, drink and make merry.
Buns, egges [*sic*], wine, lamb.
Blessing Prayer and ritual of departure.

The ritual went to plan. Before the celebration we had meditated around the cauldron on the significance of the ritual and danced around it, thus raising energy. Someone said that they saw and felt a very powerful yellow, red and blue cone of energy. They said that they could physically see it at ground level as a yellow glow, before it became weaker as it rose up and left.

Before the circle was closed a goblet of wine and some fruit were shared. Peter put quarters of chicken on the fire (in the fireplace) to cook. The time after the ritual work is done and before the circle is closed is a time for catching up on how people are feeling; it is also a time for discussion on magic. This was no exception, and a lively discussion on the power dynamics of magic circles took place. After everyone had eaten their fill, the quarters were thanked and the circle was closed. We slowly gathered our belongings and made our farewells. By the time I had taken Ken and Rachel home it was 5.30 a.m. and dawn was breaking.

This ritual, like the previous one performed by Sarah and Phil's coven, was a celebration of the Vernal or Spring Equinox – when light and dark are in equal balance, but when dark is giving way to light. The aim of these rituals is to link the witch with the seasons of the year and life processes – both internal and in wider nature. Both covens emphasized the opposing qualities of femininity and masculinity (Priestess was paired with Priest) and of light and dark in association with nature. In the first example the ritual was conducted outside 'in nature', and in the second frequent reference was made to nature 'sun', 'star', 'music of the hills', 'signs of the old stones' etc. Thus both rituals symbolized light and darkness, but this was expressed in different ways. Sarah and Phil's ritual was focused on the white and black cubes: on meditation on the cubes' and invocations, resulting finally in the cubes being ceremoniously bound. In the second ritual with Ken's coven the light and dark aspects were embodied by the performers, interwoven with the sexual symbolism of cauldron and stang. The central part of this rite was the invocation of the Light of Life by Priestess Vanessa into Peter as Priest: 'Thee we invoke, O Light of Life!'

Wicca and Invocation

Magical practices are founded on the notion of spiritual transformation. This was a central part of the Greek Mysteries, the purpose of which was to bring initiates into contact with otherworldly powers (D'Alviella 1981:33–6). In modern witch-craft there is great emphasis on the goddess. Witchcraft specifically focuses on a female deity as Goddess, who is usually seen (following Robert Graves's *The White Goddess*) as Triple. She is the creatrix of the universe, she has the primacy – although in the development of Gardnerian Wicca this was not always so (see the following section). In wiccan ritual the high priestess represents the Goddess; she creates the sacred space in the circle, and during an invocation the Goddess may be 'drawn down' into the body of the high priestess. Vivianne Crowley, a wiccan High Priestess, describes invocation as:

> . . . a process by which the Goddess or God will temporarily incarnate in the body of a selected worshipper – a priestess if the deity is the Goddess, a priest if the deity is the God. Both Goddess and God may be invoked in the rites, but traditionally only the Goddess speaks what is known as a *charge*. A charge is a ritual utterance that conveys a message from the deity to the worshipper (1993:133, italics in original).

The essence of the charge is that deity is incarnated in the body of the Priestess and that she becomes a 'secret door' of initiation (for men) through sexuality. The charge has been re-written many times, most notably by Doreen Valiente, Gardner's High Priestess, in an attempt to remove parts written by Aleister Crowley (Kelly 1991:102; for a feminist version see Starhawk 1989[1979]:76–7).

The doctrine that God can be incarnated in human form as Avatar is, according to Aldous Huxley, common to Hinduism, Mahayana Buddhism, Christianity and the Sufis (1994[1946]:59). However, usually only special people, and, in the case of Christianity, only Jesus as one unique person, incarnate deity. The incarnation of deities may be said to perform two functions: firstly, that of *bodily* communica-tion with the otherworld; and secondly, the assumption of divine otherworldly attributes and authority (I return to this aspect in Chapter 6).

In a rite called Drawing Down the Moon (see Figures 5–12) the High Priest 'using his male polarity to call forth the divine essence in her female polarity' (Farrar and Farrar 1991:67) enables the high priestess to incarnate the Goddess:

> I invoke thee and beseech Thee, O Mighty Mother of all life and fertility. 'By seed and root, by stem and bud, by leaf and flower and fruit, by Life and Love do I invoke Thee to descend into the body of thy servant and High Priestess (Kelly 1991:52).

It is significant that the words of this invocation are drawn from Aleister Crowley's Gnostic Mass, an aspect to which I shall return below. The complementary ritual of Drawing Down the Sun into the high priest occurs less often, because 'Wicca is a Goddess-oriented religion which lays particular stress on the "gift of the Goddess", women's intuitive and psychic faculties' (Farrars and Farrar 1991:68). 'By temporarily incarnating the deity in rites through the practice of invocation, the wiccan high priestess in not considered to be possessed (that is manifesting an external entity), but to be manifesting her own essential nature' (Vivianne Crowley 1993:137).

Michelle, a music student and singer in her mid-twenties, is undergoing a wiccan training, and is shortly to be initiated by Ken. She has been training with him on a regular basis for quite some time. According to Ken, there is a considerable variety of provision of training in the Craft. Some candidates undergo a year and a day's full training before initiation, while some covens offer initiation straight away, followed by training. Michelle explains how she feels when the moon is drawn down into her:

> Drawing down the Moon, for me, is an ever-changing experience each time it is performed – depending very much on the intensity of the atmosphere created during the circle-casting ritual prior to it, as well as my inner 'state' of being at that specific time/day.
>
> I have been aware of varying and differing sensations – these sometimes are quite personally significant, at other times less so felt, but always connected with or to some kind of 'transformation'. A warm tingling feeling, especially in my arms, is typically experienced – also, a change of voice as regards both pitch and tone, it becoming somewhat deeper and richer – a sense of something greater . . . always . . .
>
> At the end of the ritual I feel inspired and energised to back into daily life with a new inner, dynamic strength. During the ritual I become very much aware of my 'woman-liness', which for me is very sensual, at times rather erotic, creative or calm but always strong with a desire to 'take action'.
>
> To me it is as if each time I do Drawing down the Moon, I draw down another aspect of myself, another possible potential hidden within the psyche, within the soul – a power I believe we all possess but have forgotten how, or are frightened, to fully access and make manifest in the world.

Figure 5. In the Drawing Down the Moon ritual the priest sweeps the circle clear of all unwanted influence

Figure 6. The priestess puts the point of her athame into the water and consecrates it saying: 'I exorcise thee, O creature of water, that thou cast out from thee all the impurities and uncleanliness of the spirit of the world of phantasm; in the name of the Goddess and the God'

Figure 7. The priest purifying the priestess with the salt water

Figure 8. The priestess invokes the powers of the South

Figure 9. The priestess stands in 'Osiris position' – wand in her right hand, scourge in her left – as the priest draws the moon down into her

Figure 10. The priestess speaks 'The Charge': '. . . I who am the beauty of the green earth; and the White Moon amongst the Stars; and the mystery of the Waters; and the desire in the heart of man. I call unto thy soul: arise and come unto me'

Figure 11. The priestess places the athame in the chalice as a ritual gesture of reversed polarity – masculinity (represented by the athame) must be balanced with femininity (symbolized by the chalice)

Figure 12. The priest draws a banishing pentagram in the East as part of closing the circle

Michelle says that this rite brings out different and hidden potential within her psyche and soul. Invocation thus taps a creativity – it is inspiring and energizing – and is seen to be a source of power and transformation.

Vanessa, the art student in the Spring Equinox ritual already described, explained the process of embodying the deity 'Bride' in the following way:

> It's Candlemas, a time for women's mysteries, and I'm doing a ritual alone. Before I invoke the deity 'Bride', I begin to meditate on her, the maiden goddess of hearth who is said to dress in white and to bring three flames: that of creative inspiration, that of purification and that of love. I begin the Charge to the goddess Bride, arms raised to the sky, and my consciousness begins to shift subtly. I suddenly sense that the room is transforming, becoming alive in some way, and that a kind of ecstasy is taking hold of me. Not a wild or violent ecstasy. Something more gentle, contained. I am opening up inside and a warm flame seems to be growing outward from my womb. I feel animated, 'tingly'. Joy! 'I am she with the golden hair, queen of the white hills, rider of the white swan, and I bring three flames . . .'; as the words of the Charge lift from my lips into the sacred circle, I am imagining myself beautiful, free, draped in a silver-white gown, my bare feet rooted in the earth, my auburn hair touching the stars, small flames in the palms of my hands. The boundary between 'me' and the 'other' realms is becoming permeable, fluid; a presence seems to be filling me and yet 'I' seem to be expanding into the charged, magical atmosphere. Suddenly the cynic that lurks in my brain cuts in: is this bullshit? Is it really happening? But I manage to silence the chattering saboteur in my head, to hold the images and to move with the spiralling flow of energy. I move the energy into a dance; my hands paint sensuous curves into the air and my feet follow a spiral path. The dance is spontaneous, an offering to the goddess for her being with me tonight . . . Much later, when the ritual is over, I feel centred, alive, still, empowered. The magic has gone, but echoes of it remain inside of me.

Vanessa said that this experience of the 'light goddess' was very powerful, but that other experiences that she had had were less so. Sometimes she just felt centred and peaceful, or had a brief image or sensation. She said that she had also had more disturbing 'dark goddess' encounters, which had brought weird or ancient presences into the circle. Vanessa told me that she thought the experience of magic and altered states of consciousness depends on locale, personal mood and the company in which it was experienced.[2] For her the experience of magic involved becoming sensitive to subtle images and energies inside and outside the self and 'maybe learning how to develop and "hold" them'. Vanessa also thought that learning to explore and value the inner worlds led to a sensitivity to external forces.

Drawing Down the Sun is sometimes used for priests after invoking the goddess but before she says the Charge. Ken explained that although the title of the process emphasized the light and the 'solar principle embodied in a male', his experience of having the sun drawn down into him bore little relationship either to the Light or to the element of fire. By contrast:

The experience for me partakes of the earth element more than anything else, accompanied by a certain brownness or even blackness of mien (from this point of view it is a fundamentally grounding act). I have been told that on occasion both my voice and my visage may change. I usually feel empowered by the whole ritual sequence leading up to and encompassing this rite.

Ken said that he was affected by sound more than by any other of the senses 'the vocabulary of the bardic incantation used in this rite acts as one key trigger to what I experience . . . both in terms of phonetics and syntax'. The actual incantation/invocation varies depending on the seasonal festival enacted; but most commonly known are the stanzas found in Dion Fortune's 'to a goat-foot god'. His experience raised deep energies, which he said were impossible to get in touch with in other situations:

My experiences seem to draw upon very deep energies (they 'stir up the bottom of the cauldron'), hard if not impossible to get in touch with in other contexts (even when they may be also ritually based). There is possibly a combination of height and depth simultaneously, resulting in a neutralizing of many distinct feeling states such as anger, pity, eroticism and so on. In a sense these are transcended temporarily.

Ken felt that there was a naturalness to embodying an aspect of divinity, deity or the Sacred which made the everyday 'secular' self(s) 'decidedly second-rate'; it was more 'normal' than his everyday personality.

There are a number of points to be made arising from the experiences reported by Michelle, Vanessa and Ken: firstly, the atmosphere created during the casting of the circle, the ritual or mediation is important in effecting a change in consciousness. Vanessa noted the importance of imagery – the qualities of the goddess – and the aim of bringing creative inspiration, purification and love; secondly, the charge – the drawing down of deity – brings a deeper shift in consciousness – an 'opening inside', a 'warm tingling feeling' (Michelle), feeling animated, 'tingly'; the boundary with other realms becomes fluid, and creates spontaneous dancing (Vanessa), voice change (Michelle and Ken), and visage change (Ken); thirdly, the experience of embodying deity is seen to make a transformation that is empowering and also sensual (Vanessa), erotic (Michelle) and grounding (Ken).

The Goddess as Gateway to the Mysteries

It is clear from the Spring Equinox ritual described above that polarity is the basis of wicca. The emphasis is on heterosexual sexual dynamics as a means of raising energy in the circle. Most wiccans I have met, while stating that sexual polarity is the basis of their magical working, argue that people must find their own ways of working magic, and that the Wiccan circle may be used as an organizing framework for alternative forms of raising energy. However, it is a generally held belief among wiccans and high magicians that homosexuals and lesbians cannot work magic. In practice, I have not met many gays or lesbians who are attracted to wicca; they often see feminist witchcraft as a more flexible medium. Gardner had based his original idea on Murray's claims that witchcraft was an ancient fertility religion in a matriarchal age (Gardner 1988[1954]:31). Man eventually became master of the cult and was identified as the god of death and hunting, but: 'because of her beauty, sweetness and goodness, man places woman, as the god placed the goddess, in the chief place, so that woman is dominant in the cult practice' ([1954] 1988:31,32) So, man gives woman power, and this is reflected in Gardner's 'Craft Laws', where the position of the High Priestess, as representative of the Goddess, is given to her by the High Priest as representative of the God:[3] 'But the Priestess should ever mind that all power comes from him. It is only lent when it is used wisely and justly. And the greatest virtue of a High Priestess is that she recognises that youth is necessary to the representative of the Goddess, so that she will retire gracefully in favour of a younger woman' (Gardner quoted in Kelly 1991:146,147). Thus the High Priestess should be young and beautiful, and rules because the God gives her the power. She must recognize that when she is too old she must relinquish her place to another, younger woman. Doreen Valiente argued against much of this sexism and later, owing to the added influence of feminism, the role of the high priestess changed to incorporate female autonomy and power (see Chapter 6).

There may be differences in philosophy or psychology to explain the wiccan ritual process, but the notion that the female is central to the whole process of witchcraft is fairly universal. The Farrars state that '. . . woman is the gateway to witchcraft, and man is her "guardian and student"'[4] (1991:169). In wicca, according to Vivianne Crowley, 'the God subordinates his intellectual world to the spiritual and intuitive world of the Goddess . . .', although the ideal is holism on the inner planes through the balancing of feminine and masculine energy (the individuation process whereby the man finds his anima and the woman her animus).[5] The individuation process takes place within the framework of the feminine circle, of which the high priestess provides: '. . . the vehicle for both herself and her male partner to rise to the heights of the Godhead, united as One . . . For a woman, the part which the man will play in the rite is that of the solar hero, the rider in the chariot and it is the chariot of her own body which she will offer in order for him to assail the heights' (1989:224).

Both high magic and wicca base their philosophy on a conception of masculine and feminine difference. The difference between masculinity and femininity is perceived to be on the outer planes, gender attributes being reversed on the inner planes. The aim is to achieve the balance of feminine and masculine elements within the individual bi-sexual magical self. In wiccan practice women are seen as closer to nature (intuition), while men are seen as closer to culture (rationality);

but through magical practice both women and men come closer to their own natures (through feminine intuition).

In wicca, sex is seen to be a manifestation of the essential polarity that pervades and activates the whole universe. The Great Rite is a ritual of male/female polarity, and may be performed symbolically, with the high priestess holding a chalice of wine while the high priest lowers his athame into the wine. This symbolism may sometimes be reversed in other rituals, with the high priestess holding the athame and the high priest the chalice to represent a reversal of polarity on the inner or psychic planes. According to the wiccans Janet and Stewart Farrar, the woman and man enacting the Great Rite are offering themselves as channels for the divine polarity on all levels from physical to spiritual 'with reverence and joy, as expressions of the God and Goddess, aspects of the Ultimate Source' (1991:49). Witchcraft embraces a spirituality that rejects dominant social values concerning the body and sexuality. The Great Rite invocation declares that the body of the woman taking part is an altar, with her womb and generative organs as its sacred focus, and reveres it as such (ibid.). Actual sexual intercourse is usually conducted in private – either after the ritual or when the rest of the coven has retired until recalled.

Thus wicca is based on sexual polarity, and men and women are deemed to have different natures, both physical and psychic; it is concerned with the development and use of 'the gift of the Goddess' – the psychic and intuitive faculties – more than 'the gift of the God' – which is seen to be the linear-logical, conscious faculties, although it is said that neither can function without the other, and 'the gift of the Goddess must be developed and exercised in both male and female witches' (Farrar and Farrar 1991:18). 'The gift of the Goddess' is women's cyclical nature based on the menstrual cycle. During menstruation she is seen to be more in touch with her shamanic powers of intuition and diffused awareness. The Farrars argue that a woman has two psychic states. Drawing on Shuttle and Redgrove's *The Wise Wound*, they claim that at ovulation a woman's body 'belongs to the race as carrier of DNA', while at menstruation 'she belongs to herself', and during this time she may be in touch with her shamanic powers. Menstruation is associated with a route to esoteric knowledge. The womb is symbolized as the witches' cauldron (as in the Celtic myth of Cerridwen) – it is seen as a source of magical power. It is a cauldron of inspiration and fertility – not only reproductive fertility, but the fertility of the imagination – the essence of magic (ibid.).

The Gnostic Mass

The emphasis on sexuality as a source of magical communication is explicit in witchcraft, and probably has its roots in Tantra (See Samuel 1998:125). The essence of Tantra is the linking of the human body to the wider cosmos through dedication to the female principle and sexuality. In Tantric cosmology the whole universe is seen as being built up, pervaded and sustained by the two forces of Shiva and Shakti. Shiva represents the constituent elements, while Shakti is the dynamic potency that causes the elements to function. Shakti's name means 'power', 'for she represents the primal energy underlying the cosmos (Walker 1982). Historically Tantra has been dedicated to the female principle of Shakti, and Tantric myths and icons depict the erotic goddess as a life-giving force. It is the Tantric power of the female principle that lies at the heart of modern witchcraft. The roots of its influence lie in the work of Aleister Crowley.[6] Crowley's Gnostic Mass is central to an understanding of modern wicca, because not only does it show clearly Crowley's rebellion against his Christian upbringing – the Gnostic Mass is an explicit reversal of the Christian mass – but it places woman as the locus of deity at the forefront of the practice. It values female sexuality as a prime moving force of the cosmos; it has also shaped much wiccan ritual, especially The Great Rite – the actual words of the wiccan Great Rite were taken from Crowley's Gnostic Mass (Kelly 1991:60–1). Gardner was particularly influenced by Crowley's Gnostic[7] Mass or ECCLESIÆ GNOSTICÆ CATHOLICÆ CANON MISSÆ (from LIBER XV O.T.O). Gardner derived his ideas for witchcraft ritual, especially the idea of woman as altar, from this particular ritual.

One witchcraft coven in north London celebrates Crowley's Gnostic Mass on a regular basis: the congregation takes communion by invitation, and I was fortunate to be invited (there is a long waiting-list). It was held at a beautiful Victorian neo-Gothic house. As I walked along the pavement on my way to the Mass the houses towered over me with a dark presence all of their own. I found the number of the house and negotiated a steep overgrown zig-zag path, which was flanked with a high retaining wall covered with ivy and creepers – the scene was set.

The door was answered by a middle-aged magician in a white gold-trimmed robe, who just nodded at me when I told him my name, and led me through to a dim candle-lit hallway that smelt heavy with incense and fresh flowers. I was shown into a back room, where about fifteen people, many of whom I recognized from other magical events, were chatting, drinking tea and exchanging information. I was shown the Order of Service of the Mass. After a while a bell summoned us to the temple. Inside the temple the atmosphere changed. The room was dominated by an enormous altar in the east, which was adorned with dark pink rear curtains and a white net veil. On the altar about twenty-four white candles were burning. The congregation sat along two sides of the room. There were five officers of the Mass: the Priest (who bears the Sacred Lance), the Priestess (who should be Virgo Intacta or 'specially dedicated to the service of the Great Order' (Crowley 1991: 346)), the Deacon (who carries The Book of the Law), and two 'children' (who were in their late twenties) – one seen as negative and one seen as positive (one bears a pitcher of water and a cellar of salt, the other a censer of fire and a casket of perfume (ibid.)). The ritual focused on the veneration of woman as Priestess of the Goddess, and, for those brought up within Christian traditions, which have usually viewed women as either madonnas or whores, this was a very moving experience. The high point of the ritual is when the Priestess, who is disrobed and sitting on the altar, says:

> But to love me is better than all things: if under the night-stars in the desert thou presently burnest mine incense before me, invoking me with a pure heart, and the Serpent flame therein, thou shalt come a little to lie in my bosom. For one kiss wilt thou then be willing to give all; but whoso gives one particle of dust shall lose all in that hour. Ye shall gather goods and store of women and spices; ye shall wear rich jewels; ye shall exceed the nations of the earth in splendour and pride; but always in the love of me, and so shall ye come to my joy. I charge you earnestly to come before me in a single robe, and covered with a rich headdress. I love you! I yearn to you! Pale or purple, veiled or voluptuous, I who am all pleasure and purple, and drunkenness of the innermost sense, desire you. Put on the wings, and arouse the coiled splendour within you: come unto me! To me! To me! Sing the rapturous love-song unto me! Burn to me perfumes! Wear to me jewels! Drink to me, for I love you! I love you! I am the blue-lidded daughter of Sunset. I am the naked brilliance of the voluptuous night-sky. To me! To me!

The Priest responds:

> O secret of secrets that art hidden in the being of all that lives, not Thee do we adore, for that which adoreth is also Thou. Thou art That, and That am I.
>
> I am the flame that burns in every heart of man, and in the core of every star. I am Life, and the giver of Life, yet therefore is the knowledge of me the knowledge of death. I am alone; there is no God where I am . . .
>
> Thou that art One, our Lord in the Universe, the Sun, our Lord in ourselves whose name is Mystery of Mystery, uttermost being whose radiance, enlightening the worlds, is also the breath that maketh every God even and Death to tremble before thee – by the Sign of Light appear thou glorious upon the throne of the Sun.
>
> Make open the path of creation and of intelligence between us and our minds. Enlighten our understanding . . .

The Priestess (now robed):

> There is no law beyond Do what thou wilt.

Afterwards the congregation, one by one, went forward to participate in the Communion of a 'cake of light' and a goblet of wine, saying 'There is no part of me that is not of the Gods.'

When the Mass ended we filed out of the temple back to the other room, where a buffet had been laid out. The congregation started socializing, and cheese, biscuits and wine were handed around. I spoke to the Priestess, who explained that they had been performing the Mass on average about four times a year since 1975. The temple was their sitting-room, which had specially portable furniture which was adapted for the temple. Then I spoke to the Priest, who described himself as an urban witch who hated the countryside. He said that his leanings were towards high magic, but that he had found himself within the witchcraft movement, and so that was his path.

This ritual was very interesting to someone such as myself with a nominal Protestant upbringing; it was unusual to find a woman and female sexuality the focus of veneration on an altar. All that is viewed as at best, inferior or at worst, contaminated in Christianity is considered sacred. I now turn to a feminist development of Gardner's original formulation of witchcraft.

The Feminist Reinterpretation of Wicca

I was introduced to my first feminist witch by a mutual friend. Her name was Greta, she ran a bookshop, and I arranged to meet here there. She was in her late thirties and she was waiting outside the bookshop enjoying the sunshine when I arrived. Honeysuckle and runnerbeans were making their way up an assortment of trelliswork on the outside of the shop. She took me through the labyrinth of the shop to a back workroom, where she showed me how she recycled old material into new multi-coloured garments. Then we went upstairs into her flat, which was crammed full of pictures of the countryside, ornaments, and plants. Greta was very active in ecological action groups and had been involved in a campaign to save some local woodland. She told me about how she had had to overcome her fear when she and other protesters formed a human barricade against enormous earth-moving machinery.

Greta told me that she was brought up by an evangelist mother, but had 'converted to Paganism' two years before at a talk by Starhawk at St James's Church, Piccadilly. She said that she had been dissatisfied with Christianity, and had been investigating Celtic Christianity; but Starhawk put her in touch with magic. The very next day, at the bookshop, a woman walked in whom she had seen in the shop before, and they talked about Starhawk. The woman said that she thought Starhawk had seemed down that day, and that she had seen her looking better. It turned out that the woman was in a coven, and she invited Greta to tea so that the other coven members could meet her. Greta joined that coven, but left shortly after because she felt that the group worked using male and female polarities in very stereotyped ways. She said that it was unthinkable for them that a woman could have the god 'drawn down into her'.

Greta joined a feminist group that was open only to women. She felt that it was wrong to exclude men, but thought that it was very difficult working with men when 'some had not got past their schoolboyness about being skyclad'. She felt uncomfortable in their presence. She said that she could not get on with Dianic (women only, often lesbian) covens either.

Her partner was a Christian, and when he first found out that she was going to a coven he did not talk to her for three days. He now feels that her engagement with magic is fine, and he reads magical literature, but is not tempted to join because 'Christianity suits him'. She invited me to join an Autumn Equinox ritual. The ritual was being held above the shop the following week.

On the due day, I met the other coven members at 9.30 p.m. as arranged. There were eight of us women, six who knew each other well, and one other newcomer besides me. We took candles, nuts, grapes, apples, corn, etc. and a large dragon cauldron upstairs to the room emptied for the purpose. Before the ritual there was a general discussion about how the ritual was to run. It was very democratic, and it was explained to me that there were no degrees and no hierarchy.

We purified ourselves with water from the Lady Well at Glastonbury (someone had brought it in a plastic bottle), and we took turns at putting our negativity and fears into the water. Then, standing up, one woman did a 'tree of life' meditation for us. We felt our roots go down to earth and our spines to the stars.[8] We held hands and sang. The quarters were opened, and I was asked if I would like to invoke the spirits of the south. This I did, building images of fiery dragons and the heat of the sun. The other quarters were opened in a similar way. We then gathered around the cauldron, and another woman read out from Barbara Walker's book of *Rituals for Women*. She told us what the Autumn Equinox was about. Then we all, in turn, talked about what the Goddess had brought for the summer and what we had done.

A pathworking was conducted, and in trance we went into a field and collected some fruit. Later, when we had returned to ordinary consciousness, we talked about the various meanings, and the significance to us, of the fruit. In my trance journey I had found a pomegranate. This is a symbol of the underworld, and so we talked about the myth of Persephone's descent to the underworld. We also talked about what the winter meant to us: the cold and the dark, the internal, a time for coming together after the openness and expansion of the summer. There was a lot of positiveness: some were looking forward to the cold, the sharp outlines of the trees. The ritual concluded with a chant to the dark goddess Hecate. Bread and wine were shared, and the circle was closed.

Feminist witchcraft is a development of Gardnerian wicca; it derives from the same source as wicca, but has important differences: firstly, there is an emphasis on healing from patriarchy; secondly, this healing concerns empowerment – a reclamation of power lost owing to patriarchy; thirdly, the focus is on sexual fluidity rather than sexual polarity (see Chapters 5 and 6). This version of witchcraft developed out of the 1960s feminist critique of existing world religions, which are seen to have suppressed, devalued and denied women's religious experience; it is a re-invention of witchcraft, created by those searching for a pre-Christian era when women were valued and powerful. The awareness that the Goddess is All and that all forms of being are One is central. This approach has important implications regarding community and notions of boundaries, separation and distinctiveness, which I examine in the following chapter. Feminist witchcraft, as a political movement against 'patriarchal domination' is based on democracy (rather than hierarchy, as in wiccan covens) and a 'fluid' sexual identity that, I shall argue, brings division into the magical subculture.

The basis of much feminist witchcraft ritual practice is about re-connecting with an idealized previous state of existence, and healing the wounds of patriarchy. The practice cannot be separated from politics – a politics of re-clamation and re-invention of lost tradition. In feminist witchcraft the Goddess is considered to be creatrix – she is primal – while her relationship with her son and lover, the god, is secondary. Leah, who was a teacher in her late 30s, explained her introduction into the Craft to me. She was around in the 1970s at the beginning of the first stirrings of feminist witchcraft in San Francisco. She had a Jewish upbringing, but could not relate to the maleness of God, and so she became an atheist. In the early 1970s she had a spiritual awakening from the effects of a hallucinogen. She saw the face of God 'except it was a woman's face'. She felt an explosion of female energy that permeated her, and she realized that she was the same as everything else. 'I felt myself dissolve into molecules. I was the steps and the trees and the stars.' This experience meant that she did not have to go on 'living a miserable life as an atheist', and from there she got involved with the Craft. She explained her views of witchcraft to me:

> . . . the fact that witchcraft is a religion based on nature worship and deity as manifested primarily, though not exclusively, as female is a political fact. The fact that our God is a Goddess, that manifestation we choose to have in front of us is various shapes of female as well as male, and that the earth is female, is political. . . .
>
> The fact that we take and look at deity as a female and we take this into ourselves is a very political act, where spirituality as a whole is still owned and defined by men . . . Women should be able to be prime movers . . . although we don't have recognition in the outside world, except as satanists, which is not what we are, that's what people equate witchcraft with. The fact that we do it all is a political act, an act of personal transformation when a woman or a man are not longer able to think of God in exclusively male terms. It changes the relation to everything, other people, the planet, everything. That small difference of saying 'she' instead of 'he' is enormous.

Leah told me that the Goddess and the God were equal – two halves of all the energy in the world – and that her religion was a nature religion: 'Our religion is a nature religion, a fertility religion. It's about fucking at its most basic, not neces-sarily about biological fucking, but its certainly about fecundity and growth and death and re-planting.'

The feminist witchcraft view does not hold with an essentialist view of male and female polarity. For feminist witches sexual identity is fluid because, following wider feminist theorizing, serious difficulties have been raised concerning the assertion of the primacy of sexual difference: not only does it marginalize other forms of difference – that of race, class and ethnicity – but it also enshrines heterosexuality as the foundation of the cosmos. The feminist Goddess is a holistic unity of interchangeability and flow, and there is no clear dividing line between femininity and masculinity. Her essence is connectedness, to the extent that some would claim that all duality, including masculinity and femininity, is the product of patriarchal social construction, and therefore inseparable from power relationships. Leah told me that men's relationship with the Goddess was as brother, son, husband but not father, 'primarily not father'. She said that:

> The Goddess is immortal, she just turns. . . . He [the god] is born and dies while she turns. He is John Barleycorn or the Great Boar who is hunted and killed, sacrificed, eaten by the people, Dionysius who is ripped apart and scattered on the fields. There's life, always life, coming out – a mystery religion. She just turns. His death is not an end
> . . .
> . . . [T]here are very strong Pagan gods – animal and vegetable gods, horned and hooved gods. These are very important for men. I have a strong sense of a kind of maleness in relation to femaleness which is not the 'new man', it's not unmanly, but absolutely masculine, not afraid of masculinity, but in a different relation to himself, comfortable with himself. The god can open up things for men like the Goddess does for women, so that they can become unstuck, so they can begin to look at their bodies and feel in a different way as men, not as de-manned men but strong, vital and playful men. I grieve deeply for men, because I'd like them to be able to discover their godself, bound up with the rhythms of life, in a profound way.

I asked Leah what sort of men were involved in feminist witchcraft. She said that it was hard to say, but that she thought that the kind of Craft which she practised was difficult and it would only attract the kind of man who was prepared to take personal risks and was not afraid of women. The Craft was 'woman-strong'. It was essential for women to be strong. She thought that strong men were also essential, but 'men must possess some humour about their own qualities, so that they did not take things too seriously'. It was 'extremely important that men did not overrule women in the circle'; it was very important that 'one doesn't impose on the other':

> I think men who go into the Craft, as I know the Craft, have to truly love women, and not be afraid of themselves, they just have to really love women and be ready to not always be the one on top. Which doesn't mean that they can't be top, just that they have to be ready also not to be.

Probably women's expectations of men in the Craft are too high and that compounds the difficulties, because any man in the Craft has got to be better than a man who isn't. . . . It's all difficult and we have to treat each other gently and with as much respect as we can manage, but I do believe that if men can find their way to the Goddess that it will do what it does for women and that is make them more whole.

Witchcraft is a magical practice that opens up the social terrain of gender relationships and also issues about sexuality, and this will be discussed further in Chapter 6. For the present, I return to the area of natural magic and the association of witchcraft with nature.

Witchcraft and Nature

Pagans Caitlín and John Matthews suggest in *The Western Way* (1985) that a native tradition underpins the Hermetic tradition. They claim that native traditions started in 'Foretime' 'in which our ancestors first began to explore the inner realms of existence' and that this same tradition, which is intuitive, earth-conscious and Goddess-oriented, is manifested in contemporary Neo-Paganism. By contrast, the later Hermetic tradition pursues knowledge and oneness with godhead as super-consciousness, and this forms the basis of the mystery teachings of contemporary magical schools (1985:3). However, it is not possible to sustain a clear separation between native and Hermetic traditions, and contemporary witchcraft is part of the stream of occult ideas stemming from Hermetic natural magic. In addition, the emphasis, since the Hermetic Order of the Golden Dawn, is on internal rather than external nature. As Taylor (1989) notes, God (in this case Goddess) is interpreted as nature finding voice within.

The focus is more on internal rather than external nature; more emphasis on the often intense and emotional relationships within the coven than on a spiritual relationship with the natural world. The idea of feeling connected to nature, feeling energy 'pulsating through blood and body' as a spiritual experience, is the very essence of witchcraft ideology. This idea of connection is formed through its antithesis – the alienation from nature, developed from the Enlightenment idea that the self is essentially rational, disembodied and solitary, and given emotional weight by Christianity. Contemporary witchcraft practices are, in many ways, a continuation of the eighteenth-century Romantic movement (which was also influenced by Hermeticism), to which the philosophy of Nature as source is essential. Romanticism affirms the individual's feelings and imagination against the emphasis on rationality. The human being is a part of a larger natural order, which is inherently harmonious. 'God, then, is to be interpreted in terms of what we see striving in nature and finding voice within ourselves' (Taylor 1989:371). When contemporary witches celebrate in the woods, they celebrate the liberation of their inner selves from the domination of the everyday world. Nature represents the antithesis of urban society and its Christian values.

Modern witchcraft is mostly practised by urban dwellers. Leah explained how through witchcraft she was able to experience the turning of the year and see the planet turn. By this she meant that she was directly in contact with the forces of nature. This was vital for her to 'be able to function in the twenty-first century'. 'We are urbanized, we don't have the luxury of stepping out into fields everyday, so we've got to create it', she said. 'Robert' (Fred Lamond), a Gardnerian witch, who was initiated into Gardner's original coven, writing in a pagan journal, explains the basic tenets of witchcraft, which he describes as 'the Old Religion':

> The Old Religion . . . is a pantheistic affirmation of the unity and divine nature of all Life, in its material and spiritual, physical and metaphysical, mortal and eternal aspects. 'As above, so below.' From the neutrons revolving around the core of the atoms in the cells of our bodies, to the planets revolving around the stars in the Universe, Life is a harmonious circular dance of wheels within wheels, symbolised by the Magic Circle in which witches perform the rites (*Quest* no. 38, June 1979).

According to witchcraft lore, the seasonal rituals draw a close parallel with the cycle of birth, death and rebirth. Robert writes that in pre-Christian pagan times the aim of rituals was to ensure fertility of the crops and herds; but today their object is to 'restore empathy with nature among city dwellers alienated from it by the industrial civilisation in which they live' (ibid.).

More recently, Robert – using his real name of Fred Lamond – noted that Pagans had 'failed to inculcate a reverence for nature, even among our members'. He claimed that an ecological consciousness had grown in the wider population, but it was not from any Pagan influence, and was due to organizations such as Friends of the Earth. Lamond asked why wiccan rituals did not make wiccans more nature-conscious, and he concluded that Golden Dawn elements were used 'unthinkingly' in a concern to 'look to the book' rather than relating to nature. He suggested taking ritual cues from the land: for example, on the south coast the water (the sea) is in the south, and so the ritual element of water should be in the south and not in the west, as in a conventional wiccan circle. Lamond concluded, 'If you want to understand nature – look at nature . . . – find a tree and meditate by it for a year'.[9]

There is no doubt that some Pagans do combine their magical practice with both conservation work and ecological action. There is a Pagan organization – the Dragon Environmental group – that aims to combine Paganism with ecological issues by working with the natural powers and tides of the earth; but the magical groups that I was involved with during my fieldwork were much more concerned with their inner spiritual transformation, and did not seem particularly interested in either environmental action or in nature in general; the emphasis was on invocation and emotional states. Two of the witchcraft groups of which I was a member did not show any interest towards nature other than as a backdrop for their rituals and imagery for their sense of connectedness. Greta's feminist group was the most active in environmental issues and political demonstration; but this was largely due to the efforts of Greta and one other member. The group eventually folded as a result of differing interests, the two most environmentally aware going on to join more overt environmental protection movements.

It appears to me that often witches, like high magicians, have a deep attachment to the ceremony and intricacies of ritual procedure. Nature, despite the ideology of connection and involvement on a practical level, is seen as a beautiful backdrop against which to practise an intense, intimate and highly emotional spiritual religion. I have witnessed little active interest in the environment. One feminist witch told me that one of her fellow coven members, who was heavily involved with television 'soaps', had told her that nature documentaries on the television were 'boring'. One wiccan, when invited to go for a walk, cried off because it was raining and he might get his feet wet: 'Can't we just visualize it?', he said. These witches had a different attitude to those involved in Pagan environmental protest. I spoke to one Pagan road protester who was part of the Teddy Bear Woods Road Protest Camp in Weymouth, Dorset, and she told me that she thought Pagan activism was very different from 'ritual witchcraft'. This view is backed up by Nick Fiddes's research[10] on the 'practical Paganism' of the road protesters, who, he argues, have little concern with formal ritual (see Greenwood, forthcoming).

High Magic and Wicca: Similarities and Differences

In this chapter, and in the previous chapter on high magic, I have concentrated on two distinct yet at the same time amalgamated blends of magical practice, both of which stem from the Hermetic tradition (for the purposes of this discussion wicca and feminist witchcraft may both be classified as witchcraft). They have many similarities, but also have a number of differences that are worth outlining here. Firstly, as I have shown, both high magic and witchcraft involve both mind and body – contrary to popular stereotypes within the subculture that high magic concerns solely the mind, while witchcraft is focused on the body: this is demonstrated by the use of sexuality as an important magical practice. A second major similarity is that both high magic and witchcraft are essentially polytheistic – they recognize a plurality of divine beings – even though in the case of high magic these are sometimes subsumed to the greater divinity represented as Light. A third similarity is that both are directly shaped by Christianity: either as an esoteric variant, as with some forms of high magic, or as a rebellion against it, as in the practice of witchcraft itself.

The differences between the two broad strands of magical practice are based on opposing cosmologies. High magic is conceptually framed around a notion of evolution based on a gnostic Fall from divinity. It is therefore cosmologically dualistic – seeking to regain oneness with divinity. By contrast, witchcraft is monistic: humans are inherently good and are a part of the cosmos, which, in turn, is divine – in short, there is no conception of The Fall or sin. A major difference between the two practices concerns attitudes to women. While much high magic emphasizes 'the feminine' as well as 'the masculine', and gender polarity is seen to be the basis of much magical working, witchcraft differs to the extent that it values and expressly exalts 'the feminine': women are seen to embody what are seen as the virtues of intuition, feeling and emotion. Moreover, they are held to be superior to men at practising magic, and are often viewed as 'gateways' for men (this notion will be examined in Chapter 6). The two practices also reflect differences in attitudes to the practicalities of working magic: witchcraft is largely practised communally, while high magic has a tradition of solitary working (although high magicians do perform many communal rituals). In relation to the otherworld, high magic places great emphasis on meditation and internalizing the Tree of Life glyph, while witchcraft is *ideologically* based in nature. This reflects the degree of formality inherent within the two practices: high magic, through the use of the Kabbalistic glyph and structured pathworkings, is more disciplined in its method of working – its negotiation with the otherworld is more controlled – and, I would suggest, less open to individual interpretation. Wiccan ritual is often enacted as a form of sacred drama, with lines to be learnt and costumes to be made, but it is also flexible within its structure – there is room for inspiration and creativity, and for the 'gods to speak through the participants'.

In the next chapter I show how magicians' notions of self-identity are constructed through an engagement with the otherworld.

Notes

1. I discussed my involvement with the group and the issue of commitment with Sarah on a number of occasions. They knew about my research and I said that witchcraft was not a practice I felt drawn to on a personal level. Sarah said that when the group wanted more commitment from me they would let me know.
2. Stanislav Grof has noted the importance of what he terms the set (the individual psyche) and the setting (the social framework in which the altered state of consciousness takes place) (see Grof 1988).
3. I am grateful to Annie Keeley for drawing my attention to this point.

4. There is an increase in interest in homosexual 'queer spirituality', and a neo-pagan new religion called the Radical Faerie Movement, which began around 1978, is gaining popularity, especially in the United States, as a parallel movement to 'women's spirituality' to challenge patriarchal notions of religion (see Adler 1986:338).

5. Vivianne Crowley's Jungian view has its detractors, although her writings are very influential. A review in *The Ley Hunter* is scathing about 'Those who chant the word 'archetype' whenever they explain the divinities they worship may one day realise with a shock that they aren't just dealing with shadows in the human psyche, but powerful Goddesses and Gods with an independent existence . . .' (Summer 1989).

6. Crowley has reversed and given positive value to the Christian interpretation of witchcraft that emphasized the female witch's diabolic sexual relationship with the Devil.

7. Gnosticism may be defined in many ways: narrowly as an ancient Christian heresy; more broadly, as an antithetical dualism between immateriality, which is good, and matter, which is evil. This dualism is also present in humans, who have a yearning to restore unity of immateriality; and finally, most broadly, the belief that humans are alienated from their true selves in the widest sense (Segal 1992:3, 4).

8. It is interesting to compare this approach to the complexities of learning the Tree of Life glyph as a high magician.

9. Pagan Federation conference, Fairfield Halls, Croydon, 21 November 1998.

10. This was an ESRC-funded project on Environmental Activism: an ethnography of ecological direct action.

–5–

Magical Identity: Healing and Power

The otherworld is a source of mystery, beguilement and enchantment. Those magicians who can handle its deep, dark and powerful secrets surround themselves with glamour and intrigue like a star-spangled cloak. This is the lure of magic. It explains its popular appeal: it gives power to the powerless and most importantly, provides a magical identity. Behind the swirling, glittery mists is the idea that communication with otherworldly reality is a healing process, a psycho-spiritual path leading to spiritual transformation. Initially the magician has to heal her- or himself by 'balancing' the microcosm to become a receptacle, a 'form', for the forces of the otherworld; the human microcosm must be balanced as a pre-requisite for channelling and obtaining power from the otherworld. The process of making contact with the 'magical self' involves various psychotherapies that are incorporated in rituals as forms of healing to restore a sense of wholeness. Thus much initial magical work is psychotherapeutic in nature: it involves techniques such as pathworking and exercises to encourage such an exploration and understanding. Lévi-Strauss was the first anthropologist to make the connection between shamanism and psychotherapy. He claimed that the shaman articulated the patient's illness into a coherent system that explained the universe. The shaman was a facilitator of a healing process – the 'shamanic complex' – that involved the sick person and the 'public' who participated in the cure. The shaman's role was to articulate the patient's illness through the shaman's experiences and group consensus. The cure represented the ordering of the psychic universe into a meaningful system (Lévi-Strauss 1979[1963]). Western magicians, like indigenous shamans, order their psychic universes by working on their own healing; and in the process their worldview is changed.

Thus the otherworld is viewed as a source of healing; it is also seen to be the locus of knowledge, wisdom and power. As was noted in Chapter 2, there is an underlying similarity between all magical practices based on personal experience of otherworldly reality, which is said to lead to a shift in consciousness: a 'finding of the true self', which is the magical self. This is often paradoxical: 'finding the self' leads to 'losing the self' in the greater whole. Losing the self enables the identification and merging with otherworldly beings. This process is seen to lead to the acquisition of a magical identity and also power.

This chapter examines the central process of healing in magical transformation. Healing forms a large part of a magician's initial work, and psychotherapies have had a profound effect on contemporary magical practices. I examine the relationship between Freud and the occult and the effect that psychotherapies have had on magical work. I suggest that magical identities are structured through a psychospiritual interaction with the otherworld, rather than constructed from social discourses of the ordinary world, as suggested by some recent works on identity formation in the social sciences. The final section of this chapter deals with the important issue of power. The acquisition of power is central to the magical process, but how it is used is often at variance with the ideal practice – that of self-transformation. I tackle some of the contradictions of magical empowerment: the use and abuse of power.

Magical Identity

According to what have often been termed postmodern perspectives on identities, the self is fragmentary and the site of multiple and potentially contradictory subjectivities – subjectivity is not singular or fixed (Foucault 1990[1976]; Butler 1990; Cornwall and Lindisfarne 1993; Moore 1994). Contemporary writings on identity focus on the issue of personal meaninglessness. This work tends to view the world in an apocalyptic light as a result of the risk of massive destruction by nuclear warfare, ecological catastrophe, the collapse of global economic mechanisms and the rise of totalitarian superstates. Personal meaninglessness is, for Anthony Giddens, a fundamental psychic problem due to the disembeddedness of social relationships from the local contexts of pre-modern society to what he calls high modern (rather than postmodern) 'post-traditional' society. In high modern societies personal meaninglessness is a fundamental psychic problem, because modernity disembeds social relationships and traditions have disappeared. In high modernity the emphasis on the self leads to the 'reflexive project of self-identity'. In this view, the self has to be constructed as part of an internal psychotherapeutic process. Reflexivity of the self is seen as a form of 'excavation', to dig out 'emotional inertia' and 'debris of the past' to reconstruct the present (Giddens 1991).

Many writers claim that identity is constructed through reflexive individual achievement, often viewed as a 'narrative', which is shaped, altered and sustained in relation to the rapidly changing circumstances of contemporary social life. Kevin Murray claims that both personal and social identities are constructed by finding stories to tell about the self. Narrative becomes a way of making sense of the world. He argues that the self is constructed from a social context; the development of a sense of self is gained only through social meanings, and the primary function of a narrative is to relate theory to experience. Identity is constructed through social practices that regulate the assignment of meaning to an individual's sense

of selfhood. Identities change corresponding to the stories that govern their representation i.e. stories make identities[1] (1989:200).

Murray's work is similar to many contemporary post-structuralist anthropological and sociological accounts of the acquisition of identity. Such writing focuses
on the construction of the subject from a multiplicity of discourses and social
practices. For example, Henrietta Moore, arguing from a post-structural feminist
position, postulates that although the dominant model of the person/self in Western
Europe is characterized as individual, rational, autonomous and unitary, many
people would find it difficult to conceive of themselves as such. She claims that
Western European culture has evolved a number of ways (she cites religious belief
and popular psychoanalysis) to deal with the fact that individuals do not necessarily
experience themselves as unitary and rational. Moore believes that an anthropologist's task is to investigate how popular discourses or folk models inform
academic discourses (Moore 1994:35).

The discourse of the magical otherworld is mythological (cf. Eliade's notion
of myth as a plan existing on a higher cosmic level (1989[1954]). The wiccan
sociologist Ken Rees has shown how myths influence newcomers into Paganism.
He argues that personal myths play a formative part in the choice that a person
makes when they choose a magical practice. Rees has examined the role of myth
in creating, structuring and legitimating contemporary Paganism. Using the term
'myth' as 'an overall controlling image incorporating beliefs, attitudes and values
which direct ways of behaving . . .', he argues that a 'seeker's personal mythology
has numerous roots within family, gender identity, social class background, ethnic
and overall cultural mileu'. Drawing on David Feinstein and Stanley Krippner,
who write that 'personal myths explain one's world to oneself, guide personal
development, provide social direction and address spiritual longings' (1989:24),
Rees argues that personal myth will inform an individual's entry to the magical
subculture. There is a continuing interaction between the myths that a seeker brings
and the myths of Paganism, 'often firmed up to have the character of an ideology'.
Some seekers may choose the myth they like the best because it appears to have
the most compatibility with their initial expectations. They may later grow out of
it or become disillusioned, or they may seek for another myth elsewhere, or develop
one for themselves. Rees notes that seekers will usually stop at the level with
which they feel most comfortable – 'where there is a degree of synchrony between
their own informing myths and the outlook of the segment which they have become
lodged in', until they become disillusioned or find a new myth. They will speak
from within the mythos and reproduce it to newcomers. Rees sees the magical
subculture as a maze, or a labyrinth or patchwork quilt through which individuals
steer (1996:18–21). While Rees's explanations of myth are useful for explaining
how people adopt a particular magical practice, the exegesis of myth remains in
the realms of the ordinary world.

Myths have an emotional power in the spheres of human action, as the anthropologist Bruce Kapferer notes: 'their logic or reasoning connects with the way human beings are already oriented within their realities'. Myth provides the framework for experiencing the world and for defining significant experience in the world, it can 'contain those principles whereby human beings engage their world and realize in experience their passage through the world' (1988:46–7). Kapferer has pointed to the emotional power of myths, and it is this aspect of myth that is important in forming a magician's sense of identity. Myth is vital to the way that humans constitute a self in the everyday world, but it also essential to the otherworld: it functions as a language of alternative reality.

Magicans' self-identities are organized around a deep internal exploration of the self through an interaction with the otherworld, and the whole magical self is shaped through an intense emotional interaction with the otherworld. Thus the magician's sense of self is not formed directly from the social world but from the process of engagement with the otherworld through magical techniques such as ritual, path-working (as the visit to the Temple of Isis) or meditation on the Kabbalistic Tree of Life as described in Chapter 3.

The anthropologist Hans Peter Duerr writes about the transition between this world, or what he terms civilization, and the otherworld, an area he designates wilderness, as part of a changing sense of self. This can be compared to an iceberg: what is ordinarily visible is merely a small section of it, the part above the water. The entire iceberg is much bigger than the area that can be viewed (1985:87). Duerr points out that the process of exploring the hidden area of the iceberg concerns the dissolving of boundaries between civilization and wilderness. This involves a changing sense of personality, as the self expands and consciousness changes; the self is felt to be interconnected and a part of a whole. Ordinary ego boundaries evaporate and are no longer identical with personhood. As Duerr reflects: 'It is not so much we fly. What happens instead is that our ordinary "ego boundaries" evaporate and so it is entirely possible that *we* suddenly encounter ourselves at places where our "everyday body", whose boundaries are no longer identical with our person, is *not* to be found' (1985:87 original emphasis). Duerr notes that the expansion of person can easily be described as flying, but that this is only seen in exotic terms when the meaning of the term is fixed by the standard situation in which it is customarily applied. An expanded application would have to designate a new and different object (1985:87).

For Western magicians learning magic means learning to explore the terrain of the otherworld in a state of trance; learning to experience the 'evaporation of ego boundaries' by 'shape-shifting'. This process leads to new ways of viewing the self magically.

Rees makes a number of observations about what he calls the 'magical personality'. Firstly, it grows through ritual, the assumption of god-forms, and working with intent via spells, etc. Secondly, it is concerned with the energies inside the self, which are subject to change with the seasons, the ageing process, illness, grief, menstruation, etc. Finally, the magical personality is not developed by 'one-off experiences' such as a Gnostic Mass or wiccan initiation (although these can act as 'doorways'), but by magical training. Rees believes that the magician must be trained: just as a person learning to meditate must practice, so a magician must devote time to training within a particular tradition. A tradition will shape other-worldly experiences. The eventual aim of this deep exploration of the self is power – to be able to channel and use powerful forces of the cosmos; but for most magicians an essential step in the process is healing the self first.

Magic as Healing

All magical practices draw heavily on notions of healing as part of their philosophy and way of working. They structure magical identities: on an individual level the magician has to work to balance and harmonize inner energies as a pre-requisite to the magical work of channelling the wider forces of the macrocosm. Working in this way is seen to be therapeutic, a way of uniting mind and body and healing the splits of mind/body and the social fragmentation resulting from the pervasive cultural acceptance of this dualism.

Most magicians, except chaos magicians, say their magical practice is a spiritual path, a route to wholeness. Magic is often viewed by practitioners as a way of healing the individual from the effects of a fragmented rationalist and materialist world – it is a re-enchantment of the self. As was argued in the Introduction, the practice of magic is also in many cases a rebellion from a culture identified with a stereotyped Christianity that divides spirit from matter and mind from body. Magical practices today offer techniques (such as the guided visualization to the Temple of Isis in the introduction) to help the magician heal the body and the psyche from the disharmony of contemporary life in a bid to restore a harmonious unity. The practice of magic is seen to help restore an original balance. Healing, or more specifically ideas about healing, forms a large part of magical ideology and practice. Magical rituals may also be seen as spaces of resistance to the ordinary everyday world where alternative identities as transformative modal states for healing, what Samuel calls shamanic mechanisms for change (1990:106–20), may be worked out.

Anthropologists have long noted rituals of resistance to dominant cultural impositions among those they study, whether they are brought about through colonization, capitalism or any other agent precipitating clashes between incompatible social worlds. Millennarian movements, cargo cults, and possession cults have formed the basis of a large anthropological literature. Evans-Pritchard noted a number of Zande magical associations that had arisen since 1900 as a reaction to 'wide and deep social change' (1985:205). Called *Mani* associations, they were formed for the practice of magic in assemblies[2] to challenge traditional patterns of behaviour, and the authority of the older men.

In another case, that of the Tshidi of Southern Tswana, studied by Jean Comaroff (1986), the oppressions of colonial expansion, European capitalism, and evangelical Protestantism were resisted by the practice of Zionist rituals of healing. Protestant evangelists failed to create a unified black Protestant church in South Africa, but instead restructured by default the native conceptual universe as a language of protest. Comaroff noted that the healing of affliction was the most pervasive metaphor of the culture of Zion. Using Christian symbols and images, Zionist rituals attempted to heal the divisions created by both Protestantism and the effects of colonialism and capitalism by the harnessing and manipulation of divine power. Not only the contradictions of living as black workers within a repressive and racist regime, but also the Christian split between spirit and matter were dissolved through the healing rituals of Zion. Comaroff notes that innovative ritual comes from a disjuncture between received categories and changing everyday experience and, in the Tshidi context, served as a symbol of a lost world of order and control.

A comparison may be drawn between Tshidi healing rituals and those practised by contemporary magicians in Western societies: both have the common theme of creating unity out of fragmentation. A magician must learn to balance the forces of the macrocosm (experienced as various spirits, deities or emanations of the Divine) within the self – the forces are thus open for manipulation and re-interpretation by the magician in the harnessing of power in the creation of unity. In high magic one of my lessons stated that 'The first study of mankind is man, and after we have gone some way to fulfilling the Mystery injunction "Know Thyself" we can with some profit investigate the other evolutions.' In the high magic scheme the purpose of humanity is to express divinity. Any transgression from this path leads to 'adverse karma' – interpreted by David Goddard as 'misapplied and unbalanced energy (1996:59). By striving for healing on the inner levels – by balancing the forces within – the magician balances the forces without: 'What is brought about upon the inner levels will automatically, by universal dynamics, seek to manifest upon the physical plane' (Goddard 1996:xviii). The same principle is used for Earth Healing Day when many Pagans charge sacred sites with healing energy. A short article on Earth Healing Day, appeared in the *Wiccan* (the magazine of the Pagan Federation) asking all Pagans to participate in a fourth Earth Healing Day on Sunday 26 April 1992. Pagans were asked to combine meditation with practical action such as joining Greenpeace or Friends of the Earth, or writing to their MPs about local environmental problems, or to 'local industrial polluters' to express their concerns. The aim of this was to change the consciousness of those in government, of industrialists and others in positions of power and influence, 'so that they awake to the urgency of our environmental crisis and *act now* to save the planet'. Magical workings were synchronized at 3 p.m. British Summer Time (2 p.m. GMT), to 'create a world-wide bond of magical energy and intent to save our Mother the Earth' (1992:5) [Number 102, Imbolc 1992]. The following Earth Healing Chant was widely circulated:

> Wake the flame inside us,
> Burn and burn below;
> Fire seed and fire feed
> And make the magic grow.
>
> Shout unto our inner selves
> Men and women all,
> Open up your inner ears
> So we may hear the call.
>
> Wake the flame inside us,
> Burn and burn below;
> Fire seed and fire feed
> And make the magic grow.
>
> The silent scream of dolphins,
> The long dark wail of earth,
> The thrashing of a poisoned sea,
> The silence of still birth.
> The creeping poison in the air,
> The acid in the rain,
> Corrosion in the soul of greed,
> The fever in the brain.
>
> Wake the flame inside us,
> Burn and burn below;
> Fire seed and fire feed
> And make the magic grow.
>
> We call to you from North and South
> To come to us this hour,
> And East and West to join us
> With your strength
>   – and might and power.

The imagery in this chant invokes the spirit of fire *within* – as the magician's energy, power and will – and powerful metaphors – such as the 'long, dark wail of earth' and the 'corrosion in the soul of greed' – to fuel action to make change to heal the earth.

In addition to healing from social ills and such generalized 'earth healing', many magicians take up individual healing and/or therapeutic work. Wendy was one such magician. She was a Priestess in her thirties who ran art, psychology and spirituality workshops. She lived in an Aladdin's Cave of a flat that was on the top floor of an end-of-terrace house in east London, reached after fighting a way up a staircase crammed with books and paintings, etc. In front of the fireplace was a sand tray with lots of small figures and models, which were used for art therapy sessions. Wendy saw magic and therapy as part of one another. She explained that the practice of magic concerned changing and influencing the unformed part of the cosmos. She said that she saw more men than women being involved with the Goddess. This was because 'they want someone to relate to, not as a sex goddess, but as a real powerful woman'. Wendy also helped organize rituals for reconsecrating the wombs of women who had been sexually abused. These rituals aim to to restore wholeness after sexual violation and focus on a cleansing and reconsecration of the woman's womb. The woman is lifted into the circle and laid on the table/altar, while a Priestess says:

> It is through the divine elements that we have invoked that all life is formed and made manifest. It is from the womb of creation that we are all born and be part of the Cycle of Creation. The earth, the stars, the universe are also all wombs from which the elements are born. It is to these in turn that we must travel to find the understanding of the divine and creative force of the universe that is within and without. By cleansing and reconsecrating ourselves, we cleanse and reconsecrate the Earth, the stars and the heavens. Through this we acknowledge the divine feminine within ourselves and within all the manifest and unmanifest worlds. To create and recreate is the task of this world: to move, to change with the coming and the changing of the seasons; to let go, to die, to move on, is all part of the cycle of the feminine within and without. Feel this cycle moving through you now and forever.

On one occasion a witchcraft coven I was working with asked me to attend a healing ritual to help a friend of someone in the group. The coven had done healing work before on this man when he had first been diagnosed as having cancer. That healing work was considered a success, but now the cancer had reappeared, and they were going to send more healing energy to him. Healing rituals are based on the notion that the energies within the human microcosm have become blocked, and 'treatment' consists of working on the physical body and on the 'etheric body' – the band of energy around a person, often termed an 'aura'. The ill person is placed in the centre of a magical circle and the healing powers of an appropriate god or goddess are invoked to join the energy raised by the group. A talisman may be prepared to strengthen, protect or ward off harmful influences.

Many magical rituals that I attended ended with sending the power raised for healing purposes, either generally for the earth, for some specific ecological disaster such as the marine oil spillage outlined in Chapter 4, or for magicians' friends or families who were ill and needed some form of healing. One witch told me that working magic in rituals was also healing for the magicians involved, that it was psychotherapeutic: 'I feel personally healed, cleansed, closer to the forces that I recognize in the world, closer to the people I work with.' Psychotherapies have been influential in the way that magicians have come to understand their practice of magic as initially a form of healing from the the fragmentation of the ordinary world, and I now examine the link between psychoanalysis and magical practices.

Psychoanalysis and Occultism: A Shared Journey

James Webb (1976) has made a study of the historical links between occultism and the origins of psychoanalysis. Webb points to the way that occult ideas were applied to medicine, showing the similarity between the healing work of medicine and mysticism. He notes how, in Freud's words, 'The psychoanalyst attempts "to restore an earlier state of things" (1976:377). Healing was based on the restoration of lost equilibrium and harmony. This view is shown particularly well in the work of the Theosophist Franz Hartmann, who in 1893 published a work on *Occult Science and Medicine* that was largely based on Paracelsan precepts. Hartmann's definition of disease shows plainly the idea of restoring the lost equilibrium: 'Disease is the disharmony which follows the disobedience to the law; the restoration consists in restoring the harmony by a return to obedience to the law of order which governs the whole' (1893:10, quoted in Webb 1976:377). Both psychoanalysis and magical practices are thus based on promoting health by means of the removal of an obstruction: disease for psychoanalysis equals disharmony for magicians. In psychoanalysis a cure consists of bringing repressions to consciousness, and similarly in magical practices the spiritual 'cure' is a deep understanding of self that is seen to be part of the process of realization of immanent divinity – the self is composed of a spiritual essence, as well as of conscious and unconscious parts.

Dion Fortune wrote in *Psychic Self-Defence* (1988) that it was her experience of psychic attack that led her to take up the study of analytical psychology, which she felt, however, did not go far enough; and so she subsequently turned to occultism. Regardie, on the other hand, attempted to integrate psychotherapeutic techniques and magic in *The Middle Pillar: a co-relation of the principles of analytical psychology and the elementary techniques of magic*. His goal was to 'erase the barrier between the conscious self and the Unconscious, and to enable the student to find within the self, the Great Self who is in reality the only Saviour he will ever have' (1991a[1938]:iv). The means to achieve this end was to link the practice of magic to psychotherapy 'For Magic places the achievement of self-awareness second in importance only to the achievement of unity with God.'[3] According to Regardie, Jung's definition of psychotherapy was that which enabled the discovery of the unconscious. He viewed analytical psychology and magic as two halves or aspects of a single technical system, whose goal is the integration of the human personality. Its aim was to 'unify the different departments and functions of man's being, to bring into operation those which previously for various reasons were latent' (Regardie 1991a[1938]:16, 17).

As Webb has argued, the occult revival and psychoanalysis shared a common core of ideas,[4] the result of which is that 'psychoanalysis, psychical research, and the more religious aspects of the Occult Revival can by no means be disentangled' (1976:347). The historical study of the unconscious made possible the development of psychoanalysis.[5] Psychoanalysis developed from Freud's study of the altered states of consciousness induced by the hypnotic trances first demonstrated by Mesmer. Webb comments that the idea of 'reaching hidden areas of the mind was implicit in any of the theories of hypnosis before Freud took over the concept and developed his own special means of analysis' (1976:356). Initially Freud, influenced by his teacher Jean-Martin Charcot, sought an explanation for hypnosis in neuro-physiology, but later abandoned this view to defend hypnosis against such a purely materialist understanding (1976:355). Freud eventually turned to dream interpretation 'as another means of penetrating the hidden areas of the unconscious' (1976: 356). Freud also changed his views regarding 'occultism', from his initial stance, when he asked Jung to make 'an unshakeable bulwark' of sexual theory against occultism, to gradually altering his attitude in favour of the supernatural (1976:368). Webb comments that Freud could not escape the influence of the 'outburst of irrationalism which followed the First World War' (ibid.).

Freud saw the need to find a 'reciprocal sympathy' between occult approaches and psychoanalysis. He was influenced by Wilhelm Fliess and the notion that all life was governed by periodic laws based on the menstrual cycle. Ideas about the significance of periodicy and numerology were a part of the occult revival of the time, and Fliess's views were in accord with the idea that all of nature was linked together. This notion can be traced back to the German late Romantic School of *Naturphilosopie*. Freud saw that occultism and psychoanalysis had both experienced the same 'contemptuous and arrogant treatment by official science'. However, Freud wanted to establish psychoanalysis independently of both science and religion, and saw problems with an alliance with occultism, 'The occultist was a believer searching for reasons to support his faith, while the analyst was essentially a child of exact science' (1976:371). Webb concludes that contact with the occult could not be avoided. Psychoanalysis embodied ideas of the Occult Revival, but 'with a considerable difference from the occult originals' because they were 'secularized versions of ideas that found their traditional expression in the language of contemporary occultism' (1976:377).

It was left to Freud's one-time pupil Carl Jung, whose theories are highly influential among contemporary magicians, to develop a more spiritual interpretation of the unconscious. While Freud's ideas are connected with and underwrite occultism in all kinds of ways, Jung's psychology has had a more direct influence on magical ideologies, primarily through the work of Dion Fortune. According to Webb, Jung took his method of interpreting the unconscious from Freud, while the content of his interpretations came from occult sources. In his *Psychology of the Unconscious* (1917) Jung describes how he let himself down into his own unconscious and carried out an extraordinary journey of exploration among its contents. Whereas Freud sought to secularize occult ideas, Webb claims that Jung was the heir to unsecularized attempts to establish healing for the mind. Jung's vision of the analyst's role was nevertheless more explicitly magical than Freud's and his direct influence on the contemporary magical subculture has been greater.

The wiccan High Priestess Vivianne Crowley combines Jungian psychology and wicca, and in a newspaper article she explains why: 'One of the things that is very important in Wicca is finding the divine within you: psychotherapy has replaced a lot of things we previously got from a religious system, but it isn't enough. A lot of people seem to be searching for something more' (*Guardian* 4–5 March 1989). Crowley combines psychotherapy with wicca, which she interprets as a journey towards the wholeness of individuation. This involves meeting and absorbing into the psyche three very powerful archetypes from the collective unconscious. These archetypes are the shadow, the anima/animus and the Wise Old Woman/Man. Archetypes are seen to recur constantly in all myths and religions, and examples are 'the Great Mother', 'the Sky Father', 'the Child', and 'the Self'. The last is 'the enduring centre which is greater than our ordinary everyday selves and which endures beyond bodily death'. Wicca works with the collective unconscious and the archetypes. The first initiation degree (there are usually three; the second and third may be taken together in Alexandrian witchcraft) is an introduction to the Goddess, which concerns an opening of the unconscious to the 'shadow' or the personal unconscious. This must be confronted at the start of the journey to 'inner Godhead' (1989:149). The first initiation also involves bringing the animus or anima into consciousness and absorbing them into a self-image: a man must find in the Goddess aspects of himself that 'society has denied'. A woman must absorb qualities that she has projected on to her animus, 'As women we must recognize that it is not the animus which is the power; nor is it some male figure in our emotional life . . . ; these qualities are ours, the qualities of the Goddess' (1989:169–70).

The second initiation concerns the descent to the collective unconscious, symbolized by Orpheus descending into the Underworld (and also the Sumerian myth, 'the Descent of Inanna') to find the self. Finally, the third initiation is the sacred marriage of the shadow, the animus and anima in individuation, and this is said to activate the archetype of the person of wisdom or true Self. Vivianne Crowley writes that at this stage 'we must make the final break with all our gurus and all our teachers and cleave only to the wisdom within' (1989:221). Individuation is equated with spiritual growth and autonomy, and is thus the ideal aim in this interpretation of wiccan practice. A somewhat different view is held by feminist witches, who explicitly work at healing the self from what is viewed as the domination of patriarchy.

Healing in Feminist Witchcraft

Feminist witchcraft offers the clearest and most distinct form of magical healing from what is described as an alienating patriarchy, and its rituals are largely those of rebellion against the dominant culture. Feminist witchcraft healing is based on a restoration of all that, in the view of the participants, has been denied by patriarchal society, seen as a universal evil. Having a history in the women's liberation movement, feminist witchcraft ideology is grounded on the connection between politics and spirituality; the politics of re-claiming the self are central to its practice. Feminist witchcraft is based on a healing from patriarchy that involves a politics seeking to effect deep internal and external change. Healing is primarily based on erotic energy, relationships and democratic community expressed through ritual. Witchcraft rituals always include a cleansing and purification by water or smoke ('smudging' from an incense stick) or a combination of the two. In feminist witchcraft this normally involves passing around the circle a goblet of salted water in which participants can leave any thought, feeling or emotion they wish to be rid of. I attended one really spectacular feminist witchcraft ritual purification in London.

It was Samhain (Halloween), which is the witches' New Year. The feminist coven in which I was a participant observer had created a magical circle in the overgrown garden of a south London Victorian mansion earmarked for demolition. The entrance to the circle was guarded by two life-size paper skeletons dangling from the branches of overhanging trees. To the north of the circle was the altar, a large table crammed with apples, pomegranates, candles and a sheep's skull. In the east, the symbolic place of re-birth, was an elaborate round sweatlodge, while to the south was an enormous pumpkin lantern. The west was dominated by a large mirror, beneath which was a water bucket full of apples (these were for apple bobbing when the ritual circle had been closed). In the centre a large bonfire blazed.

The ritual centred on the purification in the sweatlodge. We were going to sweat out all our fears and bad feelings, and would thus face the new year feeling refreshed and positive. The entrance to the sweatlodge was a tunnel, and we each crawled into the darkness within, one by one. Those who were claustrophobic went last so that they could get out quickly if necessary. We sat in the near-total darkness, the shadows from the fire outside making patterns on the domed roof as we huddled together 'like piglets in the womb of the Mother'. We sweated out our anger and negativity, we howled and screamed out our feelings. It was a very elemental sensation of direct contact with the dark and the earth: we had re-established our connection with Her.

After about an hour we crawled out, one by one, making babyish noises to symbolize our rebirth, to be hosed down by icy water before being dried by the heat of the bonfire. Then the circle was opened by lighting the candles in the four quarters, invoking their spirits: of air in the east, fire in the south, water in the west and earth in the north. The Goddess and God were invoked and the ritual proceeded with singing, drumming and dancing. At one point we each left the circle in turn via a special gateway made of looped and tied twigs to gaze into the scrying mirror in the west. Who was I? What did I want in the future? What did I want to leave in the past? After a few minutes I faced the circle and shouted my name, which might be my magical name. Then I shouted everything that I was leaving in the old year. The coven shouted my name as I re-entered the circle and I jumped over the fire, to symbolize the new start to my life. I was greeted with hugs from the coven members.

Feminist witchcraft rituals function as a form of resistance to mainstream patriarchal culture. The ritual space is where the feminine is specifically valued and women are healed from the negative views of themselves that are a part of patriarchy's conditioning. The symbolism of femaleness *par excellence* is menstrual blood. Menstruation is said to be women's innate route to their shamanic deeper magical selves, and it is widely believed that women generally have a greater shamanic ability and that their experience is deeper because of their periodic rhythms; and I shall return to this aspect of magic in Chapter 6. 'Dark moon' rituals have become increasingly popular for women to celebrate and reclaim what is seen to be the most feminine part of themselves. The dark moon is seen to the 'time in the lunar cycle when the old moon dies and is then reborn again' (*Matriarchy Research and Reclaim Network Newsletter* no. 63, Beltaine 1990). The moon is often viewed as feminine in the way that it waxes and wanes, and female witches identify with its rhythms. It is a time 'to come together in a sanctified women's space' (ibid.). Rituals start with purification. Post-menstrual and non-menstruating women are also included and find it an 'invaluable way' to mark their lunar rhythms. One feminist group celebrated dark moon rituals by always wearing red clothing. On one occasion they stripped and covered themselves with red ochre to represent the blood of menstruation in a celebration of their synchronicity with the moon, menstruation and their foremothers. Menstruation, for long the subject of taboo, has become a powerful symbol of female magical identity associated with a re-connection with innate power of women.

Part of the feminist reclamation of the self regards re-identification as sexual beings (see Chapter 5). The Dianic feminist witch Z. Budapest claims that patriarchy has forcibly suppressed 'woman's basic orgasmic nature', and she sees this as a perversion representing a 'sexual glorification of the male, without the influence of the Great Mother' (1990:98). Healing thus involves discovering a 'healthy sexuality'. The healing power of sex is seen to release energies to the organism, which are taken as proof of having tapped a Divine source of energy. Resistance to sexual practices is explained by conditioning and oppression, both of which are viewed as mores of 'anti-nature, anti-woman religious forces associated with bloodshed, shame and guilt' (1990:100). The healing power of sex is also central to Starhawk's view of witchcraft. In *Anything that Moves*, a magazine for bisexuals, she links immanent earth-based religions with a sacred understanding of erotic energy as healing energy, claiming that 'the goddess traditions in particular of Europe and the Middle East were very rooted in an understanding of erotic energy as healing energy, as life force energy' (*Issue #7*, 1994). Thus both Budapest and Starhawk draw on the ideas of the erotic, which are often described as the sacred marriage, the 'mystical marriage' or *hieros gamos*. This concept was widely known in oriental and Graeco-Roman antiquity, and is described as the 'Human experience of earthly marriage and the individual's own awareness of what is godly and divine interact[ing] with one another' (G. Wehr 1990:15).

The version of feminist witchcraft propounded by Starhawk arose from the drugs recovery movement associated with Alcoholics Anonymous (AA) in San Francisco. AA provides a Twelve Step programme aimed at admitting powerlessness over alcohol and finding through the process a form of spiritual awakening: giving up the illusions of control that alcohol gives is seen to bring real liberation and strength. Starhawk develops her own system of power reclamation based on immanence and power within; admissions of personal powerlessness are the bedrock upon which power within is built (1990:165–7). Feminist witchcraft ritual is a way of becoming 'unpossessed' from patriarchy and functions to remind the witch that power is immanent: the 'knowledge of how to become possessed is also the knowledge of how to become unpossessed' (1990:96). Feminist witchcraft is primarily concerned with psychotherapeutic empowerment fashioned on the AA model.

An important aspect of feminist witchcraft healing from patriarchy concerns the creation of community through ritual – finding a space apart from social alienation. Patriarchal culture is held to be responsible for social alienation and fragmentation of the self – the cause of internal pain. Starhawk notes, 'We are creatures of situations; we must create situations in which we can be healed' (1990:97). The 'situations' for healing are the witchcraft ritual. Healing involves coming to understand the way that domination has been internalized through a psychotherapeutic reclamation of a sense of wholeness. The notion of wholeness of body is reflected in the idea of social wholeness – of the community of the witchcraft coven. The issue of community raises contradictions between the ideal of community and the actual practice, and I shall discuss the implications of this in the next section, on power.

An example of healing is provided by a feminist witchcraft workshop in north London (held in a therapy centre), run by Cybele, a well-known witch from Starhawk's Reclaiming Collective in San Francisco. This dealt with issues of class and sexuality as a form of healing. Cybele had been flown in by special demand to run two workshops – one was an introduction for men and women, and the other was for women only. The weekend I experienced was for women only and was attended by twenty six-women (all of whom where white); four had come from Germany (plus one German woman who lived in London) and one from France. On the Friday evening, after setting up a ritual circle (in one of the rooms used for therapy) and invoking Inanna and Iris, we descended to the Underworld by walking slowly in trance around the Cauldron of Inspiration in the centre accompanied by a slow drum beat. As we went down through the seven gates we shed everything that we did not need or want.

When we had descended to the lowest gate we turned to the centre of the circle and came face to face with the Cauldron of Inspiration. We stared into the Cauldron and the flames that were burning in it. Through the descent we had come to the deepest part of ourselves – our source. We chanted songs that compared the deepness of ourselves with the deepness of the Cauldron – deep to deep. We were connected to the source – source to source. As we slowly turned away to begin our ascent we were reminded that the Cauldron was still there within us. On the ascent we came up through the seven gates and brought one question, one challenge and one gift with us. After choosing one we split into small groups based on the chakra colour of the gate from which the question, challenge or gift had come. These questions, challenges or gifts were to be important on the final day, because they were the basis of making spells to bring them fully into consciousness and imbue them with our collective magical will.

The overall feeling of the workshop was one of connection and community. The second day was devoted to working on issues of class and sexuality (issues which divide). In the morning, again after casting a ritual circle, we individually went to the centre of the circle to join hands with other women on the shared basis of sexuality (heterosexual, lesbian, bisexual or celibate), or class (upper, middle, working), nationality (American, German, British, Welsh, English, etc.) and such common areas as motherhood and employment. In the morning we concentrated on class and divided into groups of upper, middle, lower-middle, and working class and one group that decided to be 'classless'. We discussed issues of the boundaries that separated us from other classes, how we felt being a part of that particular class, how we were alike, how much diversity or difference we included, how we supported each other and what we projected, either negative or positive, on to other groups. The same issues were discussed in groups composed on the basis of sexual identity. At the completion of the discussion each member of the group took a piece of paper and wrote something that they wanted to transcend. The piece of paper was tied up with red wool and thrown into the empty Cauldron of Relationship and Community. At the end of the two discussion groups each participant took out two pieces of paper from the Cauldron and carried them around for the rest of the day. At the end of the day the pieces of paper were opened and read by each woman, meditated on, and then thrown into the flames of the Cauldron to be consumed by the spirit of Kali.

Differences between the groups were dealt with magically; issues were brought to consciousness by each individual. Each woman wrote her most negative feeling or thought. These were passed on to two other women to be meditated on before being consumed by the flames – in trust to the work of the Goddess. Transformation happened magically rather than intellectually, change being seen to come from within on a deep rather than on a cerebral level. The whole emphasis of the weekend was on the healing of community. This was symbolized by passing around the ritual circle a large ball of red wool. Each woman took some to wind around her wrist or to keep in some special place to remind her that she was a part of the whole.

Magic and the Acquisition of Power

The ideal practice of magic unifies a 'healed' and 'balanced' subject (the micro-cosm) with the energies of the cosmos. The magician must learn to understand the forces and energies within her or his body, to harness them, to channel and direct them by the magical will – this is a central component of all magical practices. The subject must be unified to recognize her/his relation to the cosmos; then she or he is in a position to be a 'powerhouse' to direct these forces. The development of the magical will has been a central defining tenet of magic since the Renaissance; the magical will is associated with power – the power to channel the forces of the macrocosm through the human body by virtue of the intention of the magician.

The individual will and the role that it plays in human life has occupied philo-sophers in the West, and, according to Brian Morris, Western conceptions of the will take Kant's notion of practical reason as an expression of moral will exercised by a rational free agent as their starting-point. The emphasis on practical reason over theoretical reason prepared the way for both Schopenhauer and Nietzsche. For Schopenhauer, in a bid to counter the ontological intellectualism of mainstream Western philosophy, the will was primarily an affirmation of the 'will to live' through the unconscious, blind urge of sexual impulse. This view influenced Nietzsche, who developed his own notion, the 'will-to-power', as an energy, the inner reality of the universe, and an impulse to impose form and structure on a chaotic reality (Morris 1991:59–79). Nietzsche's 'will-to-power' expresses the essential principle of magical philosophy: that all of life is energy, and that the magician as microcosm channels macrocosmic power by the use of the will.

To Nietzsche power was central. And so it is to Western magicians. Many magicians beocome involved in magic because they lack power. Take, for example, Chrys, whom I met at a high magic conference and ritual. Chrys is in his early thirties, and by his appearance there is no mistaking that he is a practitioner of the 'occult arts'. He was a practising wiccan, but had been advised to take up high magic to further his spiritual development. Chrys had had a rather solitary child-hood. He was educated at a Catholic boarding school, where 'no one took much notice of him'. His bedroom at school looked out over Pendle Forest, which is famous in folklore for its company of witches (Hole 1986). Chrys developed a fascination with the witches, and used to lie in bed imagining what they got up to on Pendle Hill. His schoolteachers tried to discourage his interest, and gave him horrific books on witches to read, hoping that this would put him off. To no avail: this only increased his interest. Chrys left before taking his examinations, because the school had written to his parents suggesting that they were wasting their money because he was not benefiting from it.

Chrys felt a failure, because his parents had high expectations of him. He drifted into various jobs. He eventually ended up working in a bank, but felt 'humiliated' by the fact that he had not had promotion and was working alongside people his junior by fifteen years. He eventually met someone in a bar who introduced him to his first witchcraft coven. This was when he remembered his childhood fantasies of the Pendle witches. Chrys turned to magic in remembrance of his childhood rebellion against his school and also as a statement to his parents – 'they were shocked when they found out', he said. Chrys had been in an industrial accident, and he told me about how he had 'died', as his body was so badly crushed. He was rushed to hospital, but was pronounced dead on arrival. He spent a few hours in the mortuary (of which he remembers nothing). During this time the members of his coven were aware that something was wrong, and went into their temple to look after his spirit until his body had recovered and his spirit could return. Luckily, the mortuary staff had not put him in a cold drawer, and, by chance, had noticed that he had started breathing again. He was moved to intensive care for 48 hours. I asked Chrys what the doctors had said. 'Nothing', he said. 'It was covered up because they could not explain it.'

One day when Chrys was crystal-gazing, following the instructions of his witchcraft high priest, the name of a god appeared to him in the crystal. He conducted some research at the local library and, by asking other magicians, found out that this was an ancient Egyptian creator god who, according to Chrys, 'taught man, at the time of His worship, dually of arts and crafts with one hand, and of insight and magic with the other'. As humans developed their own ideas they stopped worshipping this god. The god 'stood aside and watched', but has now chosen to return and 'special people have been chosen by Him to undertake certain duties . . .'.

Three weeks later the god spoke to Chrys through the mediumship of Chrys's wiccan high priest. The god told Chrys that he had granted him the use of the power of His name to channel lost esoteric information – this was to be called the 'Great Work'. Two years later Chrys's high priest introduced him to some magicians in America who were 'channelling space information'. On one occasion they performed a ritual together. A crystal altar was set up and the Lesser Banishing Ritual of the Pentagram was performed. According to Chrys, after a short while they went into trance: 'a sphere now, electric blue it expands further till at last the room is within the sphere, living light surrounding all, and from the crystals a vertical shaft of purple light ascends . . . if we would in our minds visualise the three of us linked by beams of light we become a antenna for the communication'. Chrys was told that there was great destruction on the planet, but the energy could be contained and directed to transform the world:

> It is possible if those who are performing the Great Work have formed a matrix of energy point-to-point around your entire world, that with their focused minds, and those of us on our side working through them and with them that this raw energy of transformation may be channelled so that the entire planetary being may vibrate upwards in evolution, rather than the pattern simply being shattered.

Chrys's life has been changed through his involvement in magic. His feelings of humiliation and 'not being good enough' have been turned into feelings of being 'special'. He explains his involvement as a 'calling'. He has adopted the god's name (he even has the god's /his name printed on business cards which he hands out). He has taken the god's power into himself, he identifies with the god and is working on healing the planetary being by using his magical will (cf Lewis 1991). However, I had the feeling that this did not entirely eradicate his lack of confidence in himself as he said to me 'I don't know why I doubt myself, the gods obviously don't.'

Magicians enter magical groups and start practising magic for different reasons. Some may be attracted to a charismatic individual and to the role of disciple; they may wish to give up responsibility for their own spiritual development and hand it over to a 'spiritual leader'. However, this is said not to be the magician's path – a magician must find her or his own divine truth within. Some want the closeness of a magical group as a substitute for kin or family relationships; others may want to change political or social structures. However, I suggest that most people become involved with magic because it is associated with the acquisition of power.

Some power is used for social ends. As I mentioned in Chapter 4, the feminist coven I joined as participant observer were the most committed to environmental issues, such as protesting against the clearing of ancient woods for road-building schemes. I spoke to Greta after I had left this witchcraft group. She said that the coven that I had joined had disbanded. Everyone had grown together in the group – it had been intense – but people had learnt, and gone on to other things. Greta herself was still involved in witchcraft, but had become more committed to the Earth First environmental movement. In 1993 the Government had roadway schemes that would destroy both Oxleas Wood and Twyford Down. Both became sites of resistance to the Government's road development plans. Greta had been very busy in the Oxleas Campaign and at Twyford Down. She told me how she had learnt to be assertive and face her fear of dealing with political confrontation regarding ecological issues. She had got on earth-moving machines to stop them, and this was how her spirituality and political action were combined. This clearly gave her a new sense of self-identity and direction.

Secrecy and Mystery

By virtue of the fact that all magical groups work with esoteric rather than exoteric knowledge, their practices are imbued with secrecy, and this implies occult power. Magic is the 'other way of knowing'; it is the hidden, that which is not readily available to the wider society. Jean La Fontaine (1977), in a study of Gisu circum-cision ritual, has argued that ritual knowledge is a charter of legitimacy for social divisions. She argues that it is the possession of secrets that separates initiates from others – the secret is the experience of the ritual itself – and that this form of knowledge is used to justify and legitimate authority. La Fontaine's work has demonstrated how political legitimacy works through knowledge.

In the contemporary Western context magical power is also based on secret knowledge, and magicians identify themselves in relation to their connection with the otherworld as self-identification and as a means of self-legitimation. A wiccan high priestess that I spoke to said that to identify oneself as an initiate is a good way of defining oneself against the dominant culture. This creates an aura of mystery – a setting apart from the ordinary world. Magicians see magic as a part of the Mystery Traditions of the ancient world, which called for the individual's use of the imagination and visualization to embody the gods in a manner similar to Avatars – as a 'down-coming', a manipulation of the divine in human form (Parrinder 1982). The issue of power involves two aspects: the first concerns the experience with the otherworld – magic is personal and often intense, involving a direct emotional relationship with deities and otherworldly beings; the second is associated with this worldly identity and status. Working magic is a way of attaining status and identity, and the image of Tolkien's Gandalf the wizard, with his flowing robes and magical staff, striding out into the unknown, is a popular and compelling one.

Magicians are very sensitive to power, and highly attuned to feeling 'the power' in others. Many claim to be able to see auras, and thereby judge a person's magical/spiritual awareness, and it is a great compliment to be called psychic. At one ritual workshop I attended there were to my reckoning at least two levels of awareness in the circle: the first a presumed 'communication with the otherworld', and the second a game of psychic one-up-personship. One woman told me that she had her psychic antennae on that day, and could pick up all sorts of information about people and the way that they were behaving. She claimed to have negotiated quite a few psychic battles until she barricaded herself in a corner, saying that she had a headache.

Magicians often refer to individuals by their magical powers and abilities to generate energies in the circle. Much of this is dependent on show and the creation of an illusion of a powerful magus – there are always people ready and willing to believe and follow such a leader. Part of the mystique comes from wearing the right clothing – choosing the right robes, jewellery and hair-style to create a powerful image. Any typical magical conference is crowded with high priestesses and high priests wearing flowing robes of all colours of the rainbow. The robes may be complemented by various accessories such as staffs, talismans or magical rings. In the ordinary world, too, many attempt to cultivate an aura of power. This, in the case of high magicians, is a part of being one of the initiated, of being specially called to work for humanity's evolution – the higher initiates have the inner confidence that they are in control of the forces. The forces are the essence of magical knowledge, and being able to understand them, control them, or work with them is the heart of magical practice. Wicca, in particular, has been defined by its critics as a religion of priests and priestesses without a congregation. Indeed, many witches have set themselves up as 'witch queens' or 'witch kings'. Two witches I spoke to said that they had arranged their handfasting (wiccan marriage ceremony) at a popular Pagan spot that was an ancient sacred site. When they arrived they proceeded with their ritual, but were interrupted by another witch who had decided that he was going to lead the ritual. When the two witches said they did not want him to officiate, they told me that he shouted that they 'must be fucking mad' if they did not want the most famous and powerful witch in the country to conduct their ritual.

The Attraction to Magic

People are often attracted to magic by their feelings of powerlessness, and see in occultism a means to become powerful. Magical practices have often been associated with evil, and magicians gain a sense of power by identifying with the ability to harness powerful forces for good or ill. Many, such as Chrys, do not want magic made 'respectable', because they enjoy its oppositional quality. Chrys told me that unfortunately witchcraft energy has become 'Christianized' and that people 'behave themselves according to the dominant culture'. He believed that Paganism was a way of living in tune with nature and that sexuality was 'a natural current', but that the association of sexuality with guilt in Christianity had affected witchcraft practice, so that sexual power was not openly celebrated, and this led to abuse. Chrys told me how he thought the wiccan Great Rite was an example of the influence of Christian moral ideas about sexuality. When a wiccan high priestess and high priest engaged in ritual sex, coven members were expected to turn their backs. Chrys thought that 'sacred sexuality' should be viewed and celebrated by coven members. He thought that there was evidence of the abuse of sex in witchcraft, and he cited an example of a high priest who had allegedly formed an inner secret circle of attractive young female witches. Initiation into this select group was via sexual intercourse with the high priest, who had a 'special power' to bestow on those who showed the 'correct attributes and attitudes'. The influence of Christianity, for Chrys, represented a degradation of Paganism's true power.

In the last few years, following the claims of certain Christian fundamentalists, magical ritual has been directly associated by outsiders with child abuse. While there is no evidence to support this claim,[6] it is my impression that there are many magicians who are survivors of abuse as children, and that this is a significant factor in their attraction to magic as a form of self-empowerment. One witch I knew gained much from the positive value given to female genitalia, which had been associated with a secret part of herself, as an empowering symbol to be widely displayed. She did artwork that invariably included a sheela-na-gig, an image of a naked squatting woman revealing her vulva, often found as part of the architecture of eleventh- and twelfth-century churches. At one feminist witchcraft ritual I attended a number of the members were discussing details of a 'Vulva Show' that they were planning. Their emphasis on female genitalia seemed to be associated with a repudiation of a negative social evaluation (see Ardener 1987), but also appeared to be connected to a reclamation of the self in association with recovering a deep sense of wholeness. However, I spoke to one woman who had decided to leave her witchcraft coven after four years because she said that it did not feel right to practise any more. She said that she had been very committed and totally absorbed by witchcraft, making her robes by hand and meditating on the ritual while she stitched, but ultimately she had misgivings about witchcraft because she did not want any part of her life to be secret, as it reminded her of being abused.

Wiccan practices open up new perspectives of power for women, since the high priestess is given authority – the Goddess speaks through her. As I have argued, according to magical ideology generally, legitimacy is self-referential and internal and, according to magical tenets, is not located in any spiritual leader as such. In wicca the high priestess may be seen as a teacher, a pointer of the way, but she is said not to be a guru-figure in whom spiritual power is located. The Goddess may be drawn down into her as representative, but, theoretically, she is no more the goddess than any other woman. In short, divinity is immanent within everyone, the difference between ordinary people and magicians being that the latter are aware of it. However, some individuals are seen to be more magically developed than others, and the initiatory system of three degrees does create a spiritual hierarchy; and, I shall argue, the phenomenon of charisma plays an important role too.

The contemporary wiccan high priestess is a symbol of the otherworld, the powerful realm of darkness and the unconscious, representing the 'other' of Christianity – the powerful, sexual woman; she is the initiator of men into the mysteries. The witchcraft coven is also an alternative kinship system, of which she sometimes is the matriarch. Una, a wiccan High Priestess, told me that 'the Craft is a family' in which everyone knows their place. One of her High Priests complained that Una was a 'wonderful lady', but she had too much power: 'she rules the roost when she should be bringing others along – handing over ritual power and becoming an elder'.

My fieldwork notes reveal an extremely ambivalent entry about Sarah, the High Priestess described in Chapter 4: 'The absolute power of the High Priestess is amazing – she has the ability to paralyse individual thought, creativity and enthusiasm; her control is absolute, stifling alternative views and ideas.' Sarah had suffered a painful and humiliating childhood, and had felt intensely powerless. She explained to me that she had been abused by her father in her childhood. When I asked why her father had behaved like this, she said that he had done it because he hated her mother – hurting the children was a way of getting back at his wife, whom he considered stupid. Sarah also told me that it was her father's second marriage, and that he had also abused (physically and sexually) his other children from his first marriage. Her parents had separated when she was thirteen, and she did not see her father any more. She said that she was always aware of the presence of evil around her due to her past, and that she wanted to explain this to me. I asked her how she dealt with this, 'By exorcising it from me, but it never goes completely', she replied. I asked her how she saw evil in the world and she said that evil was there as a force, but people did have a choice – they had the capacity for good and evil. However, there was an innate tendency in some people to use evil for the development of ego. She believed that causing pain was a way of expressing power over others. This explained much about Sarah's behaviour in the coven, and I felt much closer to her at this point. However, I also felt, as a member of her circle, that the rituals were essentially about her empowerment. In one ritual she stopped the proceedings because she said the psychic energies were not right, and she questioned us individually on our visualizations to find out who the offending person was before we were allowed to continue. She finally located the 'trouble' with Ben, and she gave him instruction about visualization. The claim to be psychic greatly enhances a magician's power, and negative sanctions, or the threat of them, can cause much distress. Sarah's expression of her power was a constant part of this group, and was demonstrated in everyday coven interaction. She routinely humiliated Phil and Ben. On one occasion they both knelt down in front of her to be reprimanded for not doing the necessary cleaning of the floor before a ritual.

Perhaps the best demonstration of Sarah's power was a Beltane (1 May) ritual planned by her some months in advance. She asked the coven members, who at this time included Frances, Mark and myself as well as Phil and Ben, to be present on trust, without knowing any details. Normally rituals are discussed beforehand, but on this occasion we were put in a position of extreme powerlessness – we had no idea of what was going to happen. The only concession was that we could leave if we felt uncomfortable (normally a person must stay with the group until the end of the ritual, although temporary exits are permissible). The week before Beltane Sarah asked me if I would bake a simnel or saffron cake for the ritual. I agreed. I bought all the ingredients, and my daughter Lauren and I made the cake.

On the evening of the ritual I arrived at Sarah and Phil's house. Sarah was still in the shower and Phil and Ben were wandering around, as all the preparations seemed to have been made. Unusually, there was no shouting this time, and all seemed quite peaceful. Sarah eventually came out of the shower wrapped in a towel. She said that they had found a new house and she appeared to be in a good mood. During this time Mark and Frances, the two other coven members, arrived. Mark spoke about how he was going on a vision quest the following week in Wales. He had been two years ago and was going again. He showed us a new drum he was making. He had started to paint the front, and the back had a polished piece of coral and a carved figure on the supports. Sarah tested it for pitch, and said that it was flat and that the skin needed to be tighter. Mark said that he had been to a talk by Jonathan Horwitz on shamanism at St James's Church, Picadilly, and how he had enjoyed it. He told me how he had been interested in shamanism since 1987, and that it was shamanism that had 'put him back together after a crisis', when he was 'out of touch, physically, emotionally and psychically'. Frances was feeling rather upset because a friend had just received an abnormal cervical smear test result. Sarah spoke about the inaugural meeting for a new high magic lodge that she had been asked to join. She was unsure why she felt that she had to go on with high magic; she felt like saying that she had had enough and formally resigning, but she was giving it a little longer.

Eventually, after Phil had showered, we got down to talking about the ritual. Sarah said that she took it as a great compliment that we were all prepared to trust her by coming to the ritual without knowing what was going to happen. She repeated that we were not bound to stay and could leave at any time if we wanted to, but could we please say that we we going and not just leave. We did a ten-minute meditation to prepare ourselves, and by this time I was really curious as to what Sarah had planned. We were told to keep to the south of the circle to leave room for Sarah and Phil. After Phil had carefully bound back the sleeves of his indigo silk ritual robe, Sarah created sacred space in 'perfect love and perfect trust'. She said that 'no one who was not meant to see would see, and that a blanket would fall over their eyes. All evil would be kept out.' The watchtowers of the quarters were invoked, and Phil as High Priest drew down the Goddess Astarte into Sarah, and Sarah, as Astarte, drew down the God Cernunnos into Phil. Phil became very excited and, brandishing his athame in his hand, leapt around the circle. All the rest of the coven members were sitting in the south staring at the single candle on the floor and chanting, drumming and clapping to raise energy.

Eventually Sarah, as Astarte, demanded of Phil, as Cernunnos, whether he would sacrifice and endure pain for her. Phil shouted loudly that he would, and dramatically flourishing his athame pressed the sharp point hard against the vein of the inside of his arm. Sarah as Astarte, with Phil by her side, athame directed to the vein in his arm, approached Ben and asked him whether he would endure pain, suffer and sacrifice for the Goddess. Ben, who was in turmoil, eventually said that he was not ready. Then Phil shouted at me, the athame still on his arm, 'Will you love and endure pain for the Goddess?' I was quite shocked by this and said that I did not know what he meant by loving and enduring pain for the Goddess. I stared into the flame of the candle and said that I knew love and that I knew pain, and that I knew that for myself. Phil appeared to be unsure about this response, but eventually said that from what I had said it was clear that I was an initiate. 'So be it', I said. Next, Phil turned to Frances and demanded an answer to the same question from her. She mumbled something about wanting to learn and would they teach her. This was seized upon by Sarah and Phil, and Frances was led to the altar and asked if she was ready. She seemed dazed and a little stunned. They proceeded to initiate her. Phil drew down the Goddess into her and by this time he was charged-up, screaming and shrieking. He eventually crashed to the floor moaning. Frances looked startled and still stunned. Sarah and Phil asked her if she wanted to say anything, to which she replied that she wanted to learn and would do her best. Sarah and Phil presented her with an elaborate silver torque which was ceremoniously placed around her neck. She then had to tell the quarters her new magical name. A goblet of wine and the simnel cake, which Lauren and I had made, was passed around while a discussion ensued.

Sarah said that she had felt terrified at certain points in the ritual. She said that the Goddess had told her to present the initiation in this way. She did not know why, and she thought that it was unethical, but had decided to follow the Goddess's wishes. Originally she was going to provide a mystery, with some 'scary bits in it', but after Thursday she knew that the ritual was to be Frances's initiation. It had to be spontaneous, 'If you know that you're going to be initiated you get caught up in the ritual, and lose the significance and the immediacy.' Apparently Sarah had told Ben that lunchtime what was going to happen and he was very unhappy about it. He said he could not be present at the ritual, and had gone to do some gardening to think about it. He went back to Sarah to talk about it and she had eventually persuaded him to be present. However, he was obviously still having difficulties with the situation, because he still thought it was unethical. After the ritual both Sarah and Phil counselled him, gave him advice and told him what he was feeling. Ben did manage to challenge them a bit, but not enough, and he ended up meekly conceding that the ritual had gone well.

A discussion took place between Frances, as initiate, and Sarah and Phil. Frances said that she had felt the power go through her to her feet and asked Sarah whether she had been thinking about the surprise initiation the preceding week. 'Only since Thursday', replied Sarah. 'Oh', replied Frances, ''cos, I've been aware of some-one telling me to face a challenge and choose a path.' 'Maybe it wasn't Sarah, but the Goddess', said Phil. This suggests the ambiguity of the High Priestess's role as mediator of the Goddess. It is often unclear when it is the High Priestess who is speaking – or in this case sending psychic 'vibes' – and when the Goddess.

Phil said to me that Tanya Luhrmann's book *Persuasions of the Witch's Craft* had scared him because he understood every word and thing that she had written about. I asked him why this was scary, and he told me that he had begun to question the powers and had lost his connection and ability to summon them. However, he had worked through it and now believed that he was stronger than before. We sat around chatting until 1.30 a.m. and then Frances, Mark and I filed out of the house and into the moonlit night to go home.

This ritual raises three related issues: firstly, Sarah's childhood feelings of power-lessness associated with her father's abuse were ritually reversed to establish and confirm Sarah's status in the group – her power as Goddess – to initiate anyone who could 'face' the challenge. To offer initiation was proof to coven members of power held by her as High Priestess, and when Ben refused initiation his child-like, novice role was confirmed. Also, Phil's comments about the effect that Tanya Luhrmann's book had on him seemed to suggest that the initiation was a demon-stration – a 'proof' – of his abilities to summon the powers (in the service of Sarah as the Goddess). Secondly, the power dynamics between Sarah, Phil and

Ben raise questions about wider gender relationships in wicca. Thirdly, the issue of charisma is central in the analysis of power in wiccan covens. After the ritual Phil was absolutely drained and exhausted (even though he had had the 'sun drawn down into him' – the male equivalent to Drawing Down the Moon); but by contrast, Sarah was energized, radiant, and at that point definitely charismatic.

The practice of wicca may involve practitioners working out childhood feelings of powerlessness; this practice improves their self-conception and this-worldly identity. Sarah's role as High Priestess in this coven gave her such power. Some tentative work is emerging that associates magical practices with people who have suffered abuse. Shelley Rabinovich, in research conducted for an MA thesis, has noted that the incidence of abuse of those engaged in Paganism in North America is high. At the *Nature Religion Today* conference held at Lancaster University in 1996 Rabinovich noted that most abuse was carried out in the nuclear family and that the frequency of abuse among her informants was so high that she had to modify her questions: 1 out of 40 persons interviewed did *not* undergo trauma. She found that many people involved in witchcraft came from abused or severely dysfunctional backgrounds, with either alcoholic parents, a history of drug dependency, or physical, sexual or emotional abuse (Rabinovich 1992). Drawing on Rabinovitch's work, Sian Reid argues that magic is a metaphor for expressing and manipulating the meaning context of life: by using spells witches heal themselves from their personal histories of trauma. She claims that magical practices confer certain psychological benefits on survivors of abuse. Abuse leaves its victims feeling powerless. But by engaging in meditation, visualization, rituals and role-play – which confer power and control as self-knowledge – it is possible, she believes, to re-integrate parts of the personality to alter consciousness and work towards change. Reid argues that magic is born out of the inner strength of 'traversing darkness' – it gives a positive value to pain as part of individual growth (1996:160–3). From my own experience of Sarah's coven I could not concur with such an optimistic view. I felt that power was abused and manipulated by Sarah and that her feminine power, justified in the name of the Goddess, was largely unconscious and appeared to be out of control. Being a member of Sarah's coven reproduced my own feelings of childhood powerlessness – of not being allowed to be who I was – and I frequently had a strong desire to wear grey – to represent my loss of self – when performing rituals with the coven. I wrote about my experience in my fieldnotes: '[Sarah] undermines people so they lack confidence, so they make mistakes, so she reprimands – so the cycle goes on. She instigates paralysis of power.' This feeling of 'paralysis of power' stands in stark contrast to the received ideology that the practice of magic makes a magician powerful and also to the argument that ritual power play is part of a psycho-spiritual transformation. In my view, this coven functioned as an alternative family or kin group that reproduced power relationships based on a notion of matriarchy – where power rested in the hands of one woman.

The high priestess's role in a wiccan coven appears to subvert women's subordination to men. There were elements of sadomasochism (S/M) in Sarah's relationship to both Phil and Ben. The sexologist Richard von Krafft-Ebing coined the terms 'sadism' and 'masochism' in 1885 as medicalized 'psychopathologies of the flesh'. Sadism was defined as a desire to humiliate, hurt, wound or destroy to create sexual pleasure. By contrast, masochism was the passive enjoyment of pain or humiliation (McClintock 1993:208–10). Ideologically, the feminine power of the high priestess is part of a wider negotiated space where feminine and masculine gender roles are enacted as theatrical psycho-spiritual transformation;[7] but in Sarah's coven it was part of a power play that empowered the high priestess over other women and men. Sarah was definitely powerful after the Beltane ritual when she had the Goddess drawn down into her – she was transformed, and could be described as charismatic.

The idea of charismatic leadership, defined as spiritual authority located within an individual, stands in opposition to the notion that magic is a spiritual path leading to a magician's finding her or his true identity within. How, then, is the operation of charisma explained within magical practice? For Weber, a charismatic leader is one possessed of specific gifts of the body and spirit that are believed to be supernatural and not accessible to everybody (Gerth and Wright Mills 1970 [1948]:245). He says that charisma has no form and no organization, and only recognizes inner determination; it is internal to the individual and outside everyday activities: 'In order to do justice to their mission, the holders of charisma, the master as well as his disciples and followers, must stand outside the ties of this world, outside of routine occupations, as well as outside the routine obligations of family life' (1970:248). Weber notes that charismatic authority is unstable, and gains legitimacy from personal strength, which must be constantly proved, not by office but by miracles, to show that the leader is sent by the gods. Thus power rests upon personal experience and faithful devotion to the charismatic leader as an emanation of the divine (1970:250). Charles Lindholm states that charisma is 'a way of talking about certain emotionally charged aspects of social interaction'. It is 'above all, a relationship, a mutual mingling of the inner selves of leader and follower'. Those attracted 'feel their personal identities lost in their worship of the charismatic other' (1990:6–7).

Sarah's charisma illustrates the tension between the ideal theory and the actual practice. The wiccan ritual of Drawing Down the Moon – of the invocation and embodying of the Goddess – became a vehicle of power for Sarah: it created the space for her to become charismatic, and this was used to reinforce her position in the coven. She 'became the Goddess', and thereby could create and indulge in her own worldview and ethics, which were conveniently sanctioned by 'cosmic law'.

Wiccan covens do not explicitly worship a charismatic other, but, as I have argued, they practise rituals that form a sacred space outside the ties of this world;

and they may also function as alternative kin groups. In short, they are apart from the routine and the everyday. The Drawing Down the Moon invocation becomes the vehicle for bringing through what Weber has termed 'gifts of the body and spirit' from the otherworld. This particular aspect of wiccan ritual offers the opportunity for the high priestess to use the energies of the group to 'bring through' divine power – she embodies the Goddess, and so becomes divine herself. In addition, her relationship with the coven members may be based on 'feminine connectedness' (I shall return to this important issue). Charismatic attraction implies a loss of personal will and identity (Lindholm 1990:35), and it could be argued that relationships based on charisma are the antithesis of a search for individual magical identity.

Ritual Power

An understanding of the forces is what makes magic so mysterious; it is what makes magic secret. As La Fontaine has pointed out, 'the line which an initiate must cross to join a secret society is an invisible boundary, that between ignorance and knowledge . . . The very existence of secrets is thus crucial for the society to define its own existence' (1985:40–1). La Fontaine argues that an initiate into Freemasonry, for example, is given a body of knowledge, told the hidden meanings of esoteric symbols, the history, and the legends; yet the main secret, which is that it facilitates business transactions, is secular. Initiation into a secret society concerns the transformation of individuals from ignorance to partial knowledge; but the overall aim of the ritual is to demonstrate the power of ritual knowledge to validate the seniority of elders and legitimize the social order (1985:179–89). She argues that, to a certain extent, the individual is used in the ritual to uphold social power, and so initiation rituals should not be seen simply as a means of changing the status of individuals (1985:104).

This raises important issues for magical practices as 'Mystery Religions' of transformation, where rituals are the means by which magicians communicate with and gain power from the otherworld. There is a clear division between high magicians and wiccans on the one hand and feminist witches on the other. The former have hierarchical structures, through which magic is worked, while the latter are deemed to be democratic. The most hierarchical is high magic. The occult school of which I was an apprentice had a clear hierarchical structure, and magical knowledge was imparted through grades. According to my introductory information, an applicant is accepted as an 'Entered Novice', and will remain so until the tenth or twelfth lesson (one lesson normally taking a fortnight). The Novice wears a black robe with a white girdle for rituals. If the supervisor's report is favourable the Novice then transfers to the First Degree, and becomes a Fellow within the 'Outer Court'. At this stage black robes with a gold girdle and the insignia of the order are worn. If the supervisor's report is not favourable, the Novice will complete a further six months of work, where the standard must improve before acceptance. If a First Degree becomes Initiated the gold girdle is exchanged for a scarlet one. If an Initiate of the First Degree progresses to the Second Degree 'Inner Court', the black robe is changed for one of deep blue with an initiation girdle and insignia. The final Third Degree is by invitation only, and its initiates are termed Counsellors, and their work is concerned with 'Planetary Evolution on all levels'.

Wicca is also based on a hierarchical structure of initiatory degrees, and although everyone is seen to be equal the high priestess is 'first among equals' (Farrar and Farrar 1991:17). The high priestess of wiccan witchcraft, in partnership with her magical working partner the high priest, is overall leader of her coven: she is the 'channel and representative of the Goddess' and can 'identify directly with the Earth Mother' (1991:18). As was noted in Chapter 4, it is normal practice for the Goddess to be drawn down into the high priestess; this gives her a commanding position of power and authority to channel the Goddess force through to the coven.

By contrast, in feminist witchcraft there is usually no explicit hierarchy, and therefore no overt role for a leader. Margot Adler writes that 'Most feminists have had strong experiences in collective decision-making without leaders or stringent rules beginning in their consciousness-raising seminars and continuing in their feminist organisations and other groups'. She says that most women who come to witchcraft from feminist movements favour a non-hierarchical, informal structure, preferring loose types of decision-making including the rotation of responsibility and leadership (1986:220–1). The emphasis is on community – both political and spiritual community; but community is not unproblematic, and I shall discuss some of the associated problems in Chapters 6 and 7.

In this chapter I have shown how the ideal aim of magical practices is focused on the notion of returning the magician – the human microcosm – to a state of wholeness so that she or he may be the locus of the powers of the cosmos. Magical practices as mystery religions are concerned with the finding of the magical self in relation to the cosmos. It is not a matter of being told the esoteric meanings of certain symbols; they have to be experienced by the individual, and, most importantly, validated by the individual from this experience. Power in magical practices is said not to be located in a sacred text or a guru figure; it is viewed as internal to the self, and gained through encounters with otherworldly reality. The prospect of the acquisition of individual power is woven into magical practices as *potentiality* – but in reality it is often elusive. Otherworldly experience is mediated by all-too-human power relationships: magical 'gurus' seek this worldly prestige and status, and magical 'disciples' search for psychological security in a spiritual leader or escape in fantasy.

In wiccan coven ritual the high priestess has explicit power by virtue of her position, and I have argued that this may becomes a model for a specifically feminine charismatic power based on relatedness. By contrast, feminist witch-craft is democratic, and there is no ritual structure for individual authority and overt power – there are no ritual elements that could specifically induce charismatic power. Power is said to be de-centralized – it is not focused or directed by one person. In the feminist witchcraft rituals in which I have participated leadership roles are usually shared – one or two people may take responsibility for arranging and planning a specific event, the work falling to others for the next occasion. For feminist witches the issues of power and politics – of resistance to patriarchy – are central, and this has a direct effect on group working. Nonetheless, there are charismatic individuals in feminist witchcraft, and some feminist witches, despite an ideology to the contrary, may be attracted to such a person and to the role of disciple. Disciples give up responsibility for their own spiritual development and hand it over to the spiritual leader or guru. As Eileen Barker notes, practices leading to dependency are found not only in obviously authoritarian groups, but also in groups that claim to offer freedom within (1992:251). It is my view that there is a hierarchy even in feminist witchcraft groups, but that it is undeclared and unspoken about, and runs counter to the dominant ideals. It has become clear from my association with a number of women who are involved in Starhawk's Reclaiming Collective in San Francisco that there are certain individuals within that collective who function as charismatics and attract an admiring following. Despite the 'power within' ideology, these followers seem to be drawn more to the role of disciple and student of a guru than to the path of 'self-revelation' as part of a Mystery Tradition. One woman explained how she had been reprimanded by her teacher and made to do certain hard tasks, which, in my view, she obviously enjoyed resisting, to make her teacher press her harder as a form of attention-seeking.

The ideal aim of self-empowerment is also undermined by the actions of would-be, and actual, leaders of magical groups. In a study of the self religions Da Free John Movement, Scientology, and Est, Rachel Kohn argues that where radical subjectivism is encouraged among followers, radical authority will be exerted by their leaders. She suggests that this obvious contradiction requires subtle, and at times, jumbled discourse by the leader. The aim in movements such as Da Free John, Scientology, and Est is self-enlightenment. The single and most important sign of that enlightenment is the leader, who has achieved it, and who thus offers the expectation that followers might experience it too. However, the leaders espouse beliefs that rule out the possibility of other individuals' attaining their enlightenment by presenting ever-receding goals beyond an endless series of challenges. In this way, Kohn argues, the leader is able to maintain supremacy without actually declaring it. Kohn records that followers felt inferior in comparison to their leaders, and 'the feeling that followers express of their own puniness next to the leader, their own failure to achieve ideal functioning compared with the ease, power and the success of their leaders' was an important observation emerging from her study of self religions (1991:36). Thus authority, legitimacy and power, while being assumed by practitioners to be located within the individual, are implicitly located within the dynamics of guru–disciple relations, the magical group or even a wider magical network, such as the Reclaiming Collective.

However, Catherine Bell argues that ritualization is a strategy for the construction of certain types of power relationships, and that participation in ritual is negotiated. Using a Foucauldian definition of power as local and relational within discursive practices, she argues that ritual involves two dimensions. Firstly, relations of power are drawn from the social world and reappropriated by the individual as experience – in other words, specific relations of dominance and subordination are generated by participants simply through participation. And secondly, she argues that power relations constituted by ritualization empower those who may appear to be controlled by them, because participation is negotiated. Those seen to be controlled by ritual authority are not simply able to resist or limit the power, they are also empowered by being participants in a relationship of power. She argues that strategies of ritualization generate forms of practice and empowerment that may articulate a 'flexible strategy of consent and resistance' in terms of an understanding of the self in relation to the social. Thus power is not located in an inflexible set of beliefs or in an individual (such as a high priestess or guru), but is found in schemes that condition experience. In other words, ritual power is a discursive practice – it is not external but contained in the body as experience. Bell argues that an individual coming to ritual brings her own history as a patchwork of compliance and resistance to the hegemonic order, and participation is negotiated (1992:221).

In practice, magicians bring their own histories to the magical experience, and this shapes how they seek to use their otherworldly experiences through a flexible strategy of consent and resistance – both to the ideology of channelling power from the otherworld and also in relation to other group members. Bell's point is that participation is negotiated, and that power is never absolute, and I concur with this. Ideologically, charisma has no place within magical practices; but in practice it may, nevertheless, form a part of the complex web of human relationships, magical philosophies and practice that together constitute the magical subculture. In the next chapter I examine the implications of gender and sexuality.

Notes

1. Murray gives four narrative types: Romance – where the hero ventures to the wilderness and encounters tests; Comedy – where people are brought together in situations of extraordinary gatherings where familiar patterns of life are

disrupted and new meanings explored; Tragedy deals with elements of the self that have been discovered but are incapable of realization in the real world; and finally, Irony is reality, which often fails to live up to expectations (1989:200).

2. Each Zande locality had a lodge with its own organization, leadership, grades, fees, rites of initiation, esoteric vocabulary and greetings. Men and women (who were excluded from traditional Zande magic) joined the association in about equal numbers. Women might hold offices, even that of leader. Evans-Pritchard notes that 'The mass of members are youths and maidens and young married couples' (1985:215).

3. In his introduction to the second edition (1970) Regardie writes:

> I had first begun to read about psychoanalysis in the writings of Freud and Jung as early as 1926. I cannot say they meant very much to me, save as intellectual stimuli of a fascinating kind. When I first wrote *The Middle Pillar*, I had just entered psychoanalytical therapy, thanks to the influence of a very dear friend. The tremendous value and importance of psychotherapy as a prelude to any serious magical training was just beginning to dawn upon me.
>
> My work with Dr. E. A. Clegg of Harley Street, and with Dr. J. K. Bendit, a Jungian of Wimpole Street in London, led me to realize the importance of psychotherapy to the beginner in mysticism and magic. In fact, thirty-five years later, in 1968, I am more strongly of the opinion than I was then. So fervently do I feel about this that since that time I have acquired some of the qualifications necessary to practice various forms of psychotherapy, particularly that of Wilhelm Reich, whose work I regard as a bridge between conventional psychotherapy and occultism . . .
>
> Today I will not so much as consider even discussion of the Great Work* with a student until he has experienced some form of psychotherapy, I care not which. There is lacking, until then, a common frame of reference, and common medium of communication (1991a[1938]:viii, ix).

* Regardie defines the Great Work as the recognition that the microcosm is part of the macrocosm, which is 'a reflection of the universe, a world within himself, ruled and governed by his own divinity (1991a:ii).

4. I am grateful to Peter Pels of the University of Amsterdam for pointing out this historical link.

5. Webb, drawing on Lancelot Law Whyte's *The Unconscious Before Freud* (1962), notes that the general conception of unconscious mental processes was conceivable (in post-Cartesian Europe) around 1700, topical around 1800, and fashionable around 1870–1880 (1976:349).

6. La Fontaine, in a report undertaken for the Department of Health to determine the extent of ritual abuse in Britain, noted that from the 84 cases studied she could find no evidence that the sexual abuse of children was part of any magical

or religious rite. Of three cases of ritual abuse, defined as sexual abuse plus allegations of ritual (1994:3), she found the ritual was secondary to the primary objective of sexual abuse and that 'self-proclaimed mystic/magical powers were used to entrap children and impress them (and also adults) with a reason for the sexual abuse, keeping the victims compliant and ensuring their silence' (1994:30).

7. S/M is seen in positive terms by some Pagans, and the connections between S/M and spirituality have been explored in the United States Pagan magazine *Green Egg*. In an article on the subject Dossie Easton and Catherine Liszt argue, using Jungian psychology, that S/M is 'consensual power exchange' and that S/M fantasies come from deep places of childhood. A characteristic of S/M is an exploration of the psyche as an 'adventure in the forbidden to explore the Jungian shadow – all that is excluded from awareness in the pursuit of transformation'. They claim that, 'When we add ritual to our S/M, performing it with spiritual intention, we can travel deeper yet . . . beyond the personal unconscious mind and into universal consciousness, or spiritual awareness' (1997:18). In this context, magical ritual may be a cathartic space of transformation of the social realm: a negotiated space where feminine and masculine gender roles are enacted as a theatrical play to work out deep-seated psychological issues in the name of goddesses and gods. In short, magical ritual may be a place where social relationships of power from ordinary reality are negotiated and transgressed. It appears to me that 'spiritual S/M' is a technique like any other magical technique, and may be used or abused. Its value as a means of personal transformation is dependent on the psychological maturity, intention and integrity of the individuals concerned.

−6−

The Return of the Goddess

Invoking the powers of the cosmos is filtered through notions of gender and sexuality. The magician is locus of the macrocosm, but otherworldly powers are shaped by this-worldly ideas and practices. Thus gender and sexuality become a contested terrain – an ideological battlefield of meanings – that affects magicians' lived relationships in both this and the otherworld. Sexuality is seen to be a 'prism' through which energies of the cosmos are focused and channelled by the individual magician. It is widely agreed among high magicians and witches alike that sexual energy is the basis of all magical power: it is seen to be the essence of life itself, a creative power that underlies everything in the cosmos and that created the cosmos. Every ritual is 'powered from the sexual centre, whether or not those working the ritual are aware of it' (Ashcroft Nowicki 1991:26). Sexual energy is also a means of accessing otherworldly reality; it is a way of achieving alternative states of consciousness.

Thus sexuality is not a free-floating self-expression; it is contained, controlled and shaped by social ideas about what is appropriate behaviour for each gender, encoded in notions of femininity and masculinity; and furthermore, dominant notions of gender are largely reproduced in magical ideology and practices. However, magical practices are unique in the way that many of them particularly value the female body and the feminine as a source of power, and this has important implications for the ways that magic is worked, as well as for gender relations. In wicca in particular the female body is largely seen as a repository for a powerful feminine sexual force, whereas in mainstream culture it has been considered dangerous and sometimes evil, but always to be controlled and contained. This re-evaluation of the feminine is often expressed by the notion of the 'return of the Goddess' – a psycho-spiritual counteraction to the 'prevailing masculine mode of Western civilization' (Harding 1982[1955]:17), which is seen to have 'repressed the magical stratum' (Whitmont 1987:196). The Jungian analyst Edward Whitmont argues that consciously accepting and honouring the femininity of the Goddess can bring a new attitude to life and existence in both men and women. This femininity establishes the value of inwardness – the 'Goddess is guardian of human interiority' and counters the materialism, destructiveness, religious nihilism and spiritual impoverishment associated with patriarchy (Whitmont 1987:vii, ix). In a similar manner, Vivianne Crowley defines Paganism as 'an affirmation of the feminine as a powerful, potent, non-passive force' which 'makes it a modern religion'.[1] Wiccan rituals in particular, in contrast to high magic, which places more emphasis on balancing feminine and masculine energies within the individual, provide a space devoted to everything associated with femininity, emotion and intuition. We can see in wicca, therefore, the most developed notions of femininity and of feminine power.

Magical ritual is the space where symbolic gender polarity and the reversal of ordinary consciousness and its association with patriarchal mainstream culture are played out. Aleister Crowley's work – particularly his Gnostic Mass, as demonstrated in Chapter 4 – has located the female as the 'gateway' to magical power, and this has been taken up by wiccan practice. However, feminist witchcraft, which developed from the second wave of feminism in the 1960s and is associated with women's reclamation of power, has challenged what it sees to be gender stereotypes. In feminist witchcraft the emphasis has changed from a focus on feminine and masculine polarity to honouring a 'flexible sexuality' – an ecstatic exploration of other ways of using sexual energy.

The work of both Aleister Crowley and Dion Fortune has been formative in shaping current ideas about magic generally and in particular magicians' views on gender and sexuality. Crowley and Fortune have offered two contrasting and gendered models of magical working, which I shall term 'the will' and 'sexual harmony' models respectively. Crowley's work exemplifies a Nietzschian version of Renaissance magic – man was the locus of the macrocosm, and could use the forces of the macrocosm having learnt to develop his magical will. In Crowley's view every man and woman is a 'star' and must develop their own will. In essence, Crowley advocates magical anarchy. Crowley's will model is a relatively straightforward masculine model, which takes no account of 'the feminine' as such, but claims sexual and magical autonomy for men and women. Dion Fortune's magical work, on the other hand, was more influenced by the Jungian aim of attaining a harmonious union of masculinity and femininity. Fortune utilizes notions of a harmony between 'feminine' and 'masculine' forces (drawing on the Jungian notion of animus and anima), whereby a woman develops her psychic masculinity and a man develops the corresponding feminine complement. Fortune suggests that there are levels of consciousness leading up to the ultimate consciousness (which is beyond notions of gender) where masculine and feminine energy polarities are reversed. Thus, a woman has an outwardly passive feminine body with inner masculine energies. However, this is reversed in the otherworld, where she is seen as active and dynamic and the initiator of men. These two models have shaped contemporary high magic and witchcraft practice and the ways that people think and practise magic.

This chapter opens to a fieldwork exploration of gender and sexuality in high magic, wicca and feminist witchcraft, which leads into an examination of gender and the magical will. A debate over sexual politics within the magical subculture concludes the chapter.

Gender and Sexuality in High Magic, Wicca and Feminist Witchcraft

Susan Palmer (1994) has identified three major gender/sexuality typologies in her study of women's roles in new religions, and these are useful in understanding attitudes to gender and sexuality in high magic, wicca and feminist witchcraft:

1. *Sex Polarity* – where women and men are seen to be unequal and different (for example in ISKON (International Society for Krishna Consciousness) women are seen to be on a lower scale of purity; while in Rajneeshism women are exalted over men).
2. *Sex Complementarity*: – women and men seen as different but equal (as in the Unification Church ('Moonies')).
3. *Sex Unity*: – tolerance of sexual ambiguity (for example the Raelians).

Contemporary high magic, wicca and feminist witchcraft together embrace all three of Palmer's typological groups. High magic shows sex complementarity although some male bias and gender stereotyping can be seen – for example an emphasis on the masculine and the light is a feature of some interpretations of the Kabbalah. In wicca, sex polarity is an underlying fundamental: women – more specifically 'the feminine' – are valued over masculinity and maleness. This is exemplified in the notion of the Goddess as ultimate ground of being. Women are viewed as naturally more in tune with their bodies and the forces of nature, owing to the manner in which Western cultures associate femininity with receptivity, emotion and intuition. But, paradoxically, sex complementarity is also a basic and fundamental working principle. As in yin and yang, men and women – as masculinity and femininity – relate harmoniously together. In contrast to wicca's sexual polarity, sex unity and tolerance of sexual ambiguity are evident in the feminist version of witchcraft. The certainties of gender polarity are questioned in the wider challenge to dominant cultural notions of femininity and masculinity. I now examine notions of sexuality and gender in more detail in all three magical practices.

High Magic

As was noted in Chapter 3, the aim of high magic is spiritual evolution, and sex is used to further the process. At a high magic workshop on 'Tantra: The Sacred Science of Sex Magic' the use of sex as the means for developing more perfect forms (bodies) for the incoming forces (energies of the cosmos) for spiritual evolution was discussed. The adept leading the workshop was David Goddard. David was in his early forties, and ran seminars on a regular basis. I felt that everyone treated him with respect because he was an adept. He was quite approachable, and answered questions openly. He explained that for high magicians all life originates from one source, and that everything in the world is sacred – sexuality is no exception. Sexuality is sacred because it is life-affirming and celebrates life in a positive way – it is one aspect of the force of life. He told a group of eleven of us that Tantric sex magic aided spiritual evolution, but was not generally spoken about outside the higher grades because it was seen to be open to abuse. This was corroborated by the fact that in my apprenticeship I had been instructed not to mix Eastern kundalini practices, i.e. Tantra, with the Western magical techniques that I was learning at that time. We were gathered in his garden; outside, some children were kicking an empty can up and down the road. There were wisteria and clematis growing up a trellis and over a wall. The beautiful plants had prompted a conversation about gardening before the seminar had started. Katerina, who was in her early sixties, talked about her garden and how pleased she was that her son was taking an interest in it, because it was 'bringing out his 'feminine' side. Another woman, in her late fifties, seemed to be a gardening expert, and offered advice on organic gardening. She had come with her husband and was a little aloof – she did not say much. He, by contrast, was quite talkative, and seemed to me to have a very pragmatic attitude to sex and magic. I recognized them from high magic rituals.

David explained that the Higher Self is incarnated to seek new experiences. The Atlanteans had made a rite out of breeding rituals in order to keep the sacred blood pure, and so rituals were concerned with conception. The Atlantean mysteries were concerned with sexual function: the masculine force was dynamic and impregnating, while the feminine form had the ability to hold power coming into it. Through evolution more perfect forms were built to hold the force. It was the Atlantean task to bring concrete mind – the ability to apprehend the abstract – to humanity's evolution to create better forms by way of a breeding programme. Similarities with the Nazis' breeding programmes were noted, but it was explained that the Atlanteans were only concerned with the priesthood, and did not force their practices on the rest of society. David explained that when people make love they create a vortex of power, which is shaped like a funnel on the inner planes, and a soul is attracted into the womb of the mother. Initiates perform magical rites because they do not want just 'anybody'. 'Like attracts like', and a soul that is tied to earth (i.e. not very spiritual) will not go up very high. By contrast, an evolved soul will go high. The magicians making love will seek to create a high vortex to catch an evolved soul. He warned that it was possible for a vortex to bend sideways, and some magicians had performed black magic to bring a non-human entity from the 'dark realms'.

During the lunch-break, I talked to Robert, who was new to magic and was examining all the books on magic in the hallway. He told me how he found it difficult to find time to read – especially when there was football on the television. I also spoke to Gwen, who was a high magic supervisor and had a lot to say on women's issues, such as re-consecration of the womb rituals. We talked about a crime and punishment programme we had watched on television, where a rapist was confronted by a woman who had been raped. She told me how she thought much therapy was superficial – in her view proper therapy took a long time.

In the afternoon, David explained that the sex force could be used and directed to magical ends, and that the use of sex for the changing of consciousness was first heard of in the 'Atlantean workings'. He said that orgasm expanded the mind to a state of 'at-oneness with the cosmos' – and this was seen as a deep bonding. The kundalini or serpent power is used in advanced magical rituals to release a 'devastating power', which, according to David, 'is never awoken except under the guidance of a teacher under controlled conditions'. He explained that the use of sex magic gives rise to the birth of thought forms on the inner level. This is seen as the birth of consciousness, and is used to create thought forms such as Dion Fortune's Moon Priestess and Moon Priest workings for the liberation of women. Sex is used as an act of worship in the celebration of a force, and it is said that 'inner plane forces' (gods and goddesses) enjoy having sex with humans as a 'supreme act of alchemy' – an enhancement of consciousness. The adept explained that sex magic could also be 'the pits'. This was because during ordinary orgasm the mind 'turns a somersault' and goes into a free state. By contrast during sex magic a thought form has to be built up before the sexual encounter, and then at orgasm the thought form is called up so that the power can be impregnated with it. 'The mind has to be snatched away from the pleasure of the orgasm and sent into the intent.' The power thus released follows the mind in whatever form the mind has made.

This sex magic was not part of my training as a student magician; but as I outlined in Chapter 3, I did learn the attributes of the various sephiroth (or spheres) of the Kabbalistic 'Tree of Life' glyph, and these were gendered. I was taught that the sephirah Chokmah represents wisdom and the great primary male force. It is concerned with all masculinity, the father, and father-figures such as employers, authority and the state. The sephirah Binah represents understanding and is associated with form, restriction and limitation; it comprises all that is latent or passive – it is the 'social unit' in which the force of action has to work. It is the Great Mother, representing death and rebirth but also the home as the domain of the feminine side of life. The aim of learning the Kabbalah is to internalize the attributes of each sephirah of the 'Tree of Life' glyph; and so although the sephiroth of Chokmah and Binah are gendered in their association with notions of masculinity and femininity, the high magician must work towards balancing all the attributes of all the sephiroth within the self, leading to a form of spiritual androgyny. As my magical supervisor pointed out: '. . . on the inner, we are neither male or female – we just are.' Thus the emphasis is on balance – balance equals wholeness. The high magician works to becoming whole – aligning the microcosm (the self) with the macrocosm (the cosmos), and thereby helping humanity to evolve spiritually. As my magical supervisor commented: 'When we are whole (or well on the way to being) we alter the whole of our race and in turn, the Universe.'

Part of my training as a high magician was based on meditating on the Chokmah and Binah, and my magical diary record of mediations read like this:

Friday 15 January 1993   11.30 – 12.00 pm
Bedroom: relaxed, breathing OK, but slight headache

Interwoven Light exercise [in this exercise a broad band of white light is imagined to be wrapped around the body starting from the feet and moving from right to left upwards] – visualised Chokmah (Jah). The grey light of the Sephirah turned sparkling silver-grey as I wound it around my body. I could feel the energy radiating in all directions – inner and outer as I felt the force of Chokmah – uncontained and unrestricted – sparkling and igniting the universe with its energy.

Saturday 16 January 1993   11.00 – 11.30pm
Bedroom: relaxed, breathing OK. Feeling emotional.

Interwoven Light exercise. Chokmah again. Strong feeling of uncontained energy, male energy. Rather frightening at first on its own without its female counterpart. Probably because I associate this with violence. The energy was sparking in all directions . . .

Realisations
This was rather scary, but if I look at the energy as 'impetus' which is ultimately contained within Binah, I can understand it better.

Sunday 17 January 1993   12.00 – 12.30pm
Bedroom: Relaxed, breathing OK, still emotional.

Interwoven Light exercise. Binah (Jehovah). This feels a very different experience – as black bands come around me I feel the containment – such a contrast to Chokmah. As they come higher and higher I start to feel panicky that I'm going to suffocate. Then I image an enormous Mother Goddess who is holding me in her arms (I am very small), I can just about breathe . . .

Monday 18 January 1993 2.00 – 2.30pm
Bedroom: Relaxed, breathing OK.

Interwoven Light exercise. Binah again. I feel drawn into the darkness, I am aware of the terrifying presence of the Great Mother, who is the Great Destroyer. I become smaller and smaller as I disappear into the infinite Darkness – the Silence is deafening and She destroys me as I disappear into a spiral of nothingness.

Realisations
This is scary and powerful stuff – The Great Mother who gives and takes life – she who gives ultimately re-claims. She is a containing force, limiting, which can also be comforting – a great paradox – again!

My training as a high magician involved a dialogue with Pearl, my magical supervisor, who responded to my meditation records. Pearl, whose mother was a student of Dion Fortune, told me that I worked strongly with the feminine, and asked me where my feminist ideas came from. I was told to let go of any belief systems such as feminism or patriarchy. When I questioned her about the gender of God she said: 'If you want to worship God in the female aspect, that's fine, but remember that male and female forces are equal – neither are superior or inferior, and the Universe has to have both to be balanced both on the macrocosm and the microcosm.'

We also had an ongoing discussion about the differences between the myth of Inanna and the sacrifice of the god. This revealed the way in which a magical training, such as the one that I underwent, constructed a certain framework for explaining and interpreting not only the experiences of meditation but also the wider world. Obviously this framework was gendered. Often associated with initiation into the Mysteries, the story of Inanna is said to be 2,000–3,000 years older than that which tells of Jesus' death, descent into hell and resurrection. According to Baring and Cashford the myth of Inanna is the oldest ritual dramatization of a lunar myth:

> . . . Inanna makes her descent into the dark realm of her sister, Ereshkigal, removing, piece by piece, the regalia of her office at each of the seven gates of the underworld. Ereshkigal fastens on Inanna 'the eye of death', and for three days she hangs like a carcass on a hook. Her faithful companion, Ninshubur – whose name means 'Queen of the East'[2] – whom she warns to go in search of help for her if she does not return, appeals to the god Enlil, then to the moon god, Nanna, and finally to Enki. Enki, the god of wisdom, reponds to her and sends two creatures to plead with Ereshkigal for Inanna's release. They find Ereshkigal in the process of giving birth. Inanna is restored to life and ascends like the moon after its three days' 'death' to assume her place once more as Queen of Heaven. But she is forced to appoint someone as a sacrifice in her place and, refusing to allow Ninshubur to be sacrificed, or her sons, she chooses her husband, Dumuzi (1993:216).

In my meditation I had conflated the myth of Inanna and the sacrifice of the god:

Wednesday 16 September 1992   11.30 – 12 noon
Bedroom: fairly relaxed, but overtired.
Meditation subject: Tiphareth sphere.

I go down through Kether, Daäth and into Tiphareth. I feel the mystery of the dying and rising god. What does this mean? What does it mean in Christianity? I feel that this is very profound and that I don't fully understand. Again, it's about the death/life cycle – the god-man who dies and is reborn. This is the same as the descent of Inanna, but does the god man do it for us – surely this denies us our experience – or do we do it 'with' him? – all these questions come to my mind.

This meditation, according to Pearl, showed my confusion about masculine and feminine roles in magical practice. The saviour, according to this magical philosophy, is always male, and I had to find and 'activate' my 'masculine self'. The saviour figure is the central symbol in Christianity because he mediates between divinity and humanity. In the Hermetic Kabbalah the sephirah Tiphareth is associated with the heart centre – sacrificing the lower self to the higher self. Pearl told me that the myth of Inanna was not the same as the sacrificed god. The myth of Inanna, she said, was about facing the dark side, a 'descent into hell to deal with the worst part of yourself'. This was seen in terms of Jungian psychology – of coming to face your shadow, your own repressed feelings and fears (Perera 1981). By contrast, she said the sacrifice of a god such as Odin, John Barley Corn, Osiris or Jesus is about a god dying 'so that we can all experience god'. She said there could be a goddess sacrificed 'but there just doesn't happen to be one', belying the fact that in this philosophy inflexible notions of gender prevent role reversal. As a high magician becomes more aware she or he becomes 'initiated into a higher level of consciousness' and sees the need to sacrifice her or his lower self for the higher. The director of my high magic school, Dolores Ashcroft Nowicki, wrote that these gender roles are clearly demonstrated by the notion that woman is the earthly image of the Great Mother Goddess, a symbol of 'the eternal and limitless outpouring of spiritual life'. Man, on the other hand, symbolizes the 'seasonal, fluctuating and limited side of earthly life'. It is the masculine role of the God to die, and the Goddess seeks him out and revitalizes him to give birth to life. Ashcroft-Nowicki claims that: 'This is the bottom line of the sacrifice myth: the Goddess conceives by the God, offers him as a sacrifice to the earth, then descends into Her own inner being to gestate his seed into the new God who, as Her son/consort, will impregnate Her and die in his turn. It is also the pattern of initiation, meaning "re-birth" . . .' (Ashcroft-Nowicki 1991:36).

The dialogue with my magical supervisor about gender roles demonstrates clearly the issue I raised in Chapter 3 about how a major part of training to become a magician is learning to interpret experiences from the otherworld through a dominant ideological framework. It is essential that this framework replaces other philosophies, outlooks and ideas. For example, when I asked Pearl about patriarchy and feminism I was told to 'let go of all my previous belief systems', because high magic is about balance and development of a higher awareness of spirituality, the aim of which is to open up, understand and balance the inner self in preparation for the transmutation of the self with the Ultimate, and thereby to aid humanity's evolution. She told me that we must be responsible for the planet and planetary beings, but if we do destroy the planet, 'like water, it will eventually find its own level', meaning that the laws of evolution will continue nonetheless, and we will 'take knowledge in the group mind and we will be given another chance to use it'. This attitude obviously takes the urgency out of any politics for change, and stands in radical contrast to the activities of feminist witches. As I have already described in Chapter 3, the more I questioned ideas and issues the more my magical progress was hindered. If I accepted the teachings without reservation and reinterpreted my worldview within this ideological framework I was rewarded with comments on how much I was improving.

Dion Fortune's main influence on magical practices was on the dynamic nature of polarity: she was the first magician to incorporate Jungian psychology into magical ideas and explicitly to give women magical power. For Fortune the sephirah Binah is associated with form and restriction, but this does not mean that femininity is viewed solely as passive: the relationship between the two aspects of polarity is active and dynamic, and women are seen to be passive on the outer planes, but are active on the inner: 'In the outer, he is the male, the lord, the giver of life. But in the inner he taketh life at her hands as she bendeth over him, he kneeling. Therefore should he worship the Great Goddess, for without Her he hath no life, and every woman is Her priestess' (1987:132).

In her magical novel *The Sea Priestess* Fortune tells of Wilfred Maxwell, a feeble asthmatic, who is taught the secrets of magic by Morgan Le Fay, the Sea Priestess. Fortune uses the medium of the novel to convey what she sees as essential magical lore. The novel explains the mystery of how what is 'dynamic in the outer planes is latent in the inner planes' (1989:160). The Sea Priestess is an initiate of the 'higher mysteries', and was herself taught by the Priest of the Moon that manifesting Life had two modes or aspects – the active, dynamic, stimulating and the latent and passive, which 'receives the stimulus and reacts to it'. He showed her 'how they changed places one with another in an endless dance, giving and receiving: accumulating force and discharging it; never still, never stabilized, ever in a state of flux and reflux as shown by the moon and the sea and the tides of life . . .' (1989:159). In turn the Sea Priestess taught Wilfred that '. . . in every being there are two aspects, the positive and the negative; the dynamic and the receptive; the male and the female . . .' (1989:16).

Thus there is gender equality on the inner magical spiritual domain; but this may be expressed using gender polarity on the outer planes and in the performance of ritual, as a high magic ritual 'The Summer Greeting Rite' shows. This ritual demonstrates the close connection between high magic and wicca and the cross-over of ideas: it is enacted around a central cauldron (a symbol of femininity borrowed from wiccan ritual) and focuses on the Lady and Lord as sacred couple. The Lady summons the feminine powers of earth and water (rivers, streams and lakes), while the Lord summons the powers of air and fire, which are identified with masculinity:

> LADY: I am She who is the summer's breath, who holds the balance of the year.
> I am She who is in the brooks set free from winter's snare.
> I am She that in dappled shade of full leafed tree will whisper:
>> Come to me!
> I am She in whose care all things have their time and place.
> I am She who replaces madcap maid with loving woman's face.
> I am She who calls to man and beast, bow before the power of love.
> Come to me!
> I am She whose warm breath feeds the earth and turns the fields to gold.
> I am She who is the core of every fairy tale that's told.
> I am She without whose love the life of men would wither all away.
>> Come to me!

Compare the role of the Lord:

> LORD: Look on me without terror, for I am strength.
> Lean on me without restraint, for I will hold thee.
> Know me for what I have always been, the Life Force of the Land.
> My Horns are the symbols of that Force, as is my upraised phallus.
> Do not turn from me in shame, but glory in my manhood
> For it but reflects your own.
> Hear my voice in the October wind and the Winter's silent snow.
> Feel my power in the surge of Spring and Summer's bounty.
> Weep not when I am cut down, for I will rise again.

Note that Binah, the Great Mother represented by Nature, is receptive: 'Come to me!', and that Chockmah, the masculine forces symbolized by the phallus, represents strength and 'the Life Force of the Land'.

Despite the strong emphasis on gender roles, high magic outside occult schools may be worked in ways that avoid sexual polarity. I spoke with Rosemary, a lesbian high magician in her thirties, about sexual identity in magic. She had been in the same high magical group for four and a half years. She explained that any member of the group could take any energy, for example, Venus (a traditionally 'feminine' sphere) or Mercury (traditionally 'masculine'), and work with it in any way they wished. She felt that high magical groups were theoretically much more flexible than wiccan covens, who, she said, tended to polarize energies. Rosemary told me that the tree of life was a very flexible system – 'the energies can be balanced in many ways' – and it was the way that it was interpreted that was stereotyped. She explained about ways that she has worked with other women using different energies: as a black goddess and a white goddess, or the spirit of earth with the spirit of sky. Energy is mediated like a pattern of eight, like yin and yang – both need each other. She also noted differences between heterosexual, gay and bisexual men, and said that their sexuality affected their magical practice, 'Bisexual and gay men are more attuned and less aggressive', she said. She also thought that, 'Magicians who work regularly in a balanced way become more hermaphroditic.'

Wicca

Contemporary wicca, as was noted in Chapter 4, is practised in small groups called covens, and attaches great importance to the aspect of family. According to Frederic Lamond, an initiate of Gardner's original coven, wicca is a family that is based on gender polarity, but with a 'special emphasis on femininity'.[3] Wicca is unique among magical practices in that the head of a coven is usually a high priestess. When a witch reaches the Third Degree, the highest level of initiation, she may 'hive off' to form her own coven. Polarity is essential to the way that witchcraft is worked. Lamond points out that all is divided into male and female, and perpetuation requires the union of both genders, and this is the basis of the nature of the powers: the primal mother and father brought earth and all its creatures into existence. This polarity is portrayed by the high priestess and the high priest in the highest wiccan sacrament of the 'Sacred Marriage' or 'Great Rite'. This ritual aims to channel what are seen as the essential male and female sexual polarity of high priest and high priestess into the cosmos in such a manner as to transcend duality.

The principles of wicca were explained to me when I went to visit a coven just outside London. The High Priestess, Una, and two High Priests lived in a small Victorian cottage that was the last house in a narrow street. I knocked on the door and was greeted warmly by the High Priest, who ushered me past the broomsticks into a small room filled with people. I was introduced amid a great commotion of getting things ready for the High Priestess to come down from the bedroom. Her sandals had to be fetched, the right tee shirt had to be found. I took a look around the room and my eye was immediately attracted to the hearth, which was bristling (and bulging) with magical paraphernalia. The Goddess Hathor nestled alongside figures of Bast (an Egyptian cat goddess) and Pan. There were cords, candles and the twin Kabbalistic pillars of Jachin and Boaz.

Eventually the High Priestess emerged, I was introduced, and then she left, accompanied by the High Priest, to have her hair done. I was left to speak to the second High Priest who was called 'Foxy'. He was between 50 and 60 years old and complained of his rheumatism. He told me how Una had taken him in after his wife had 'turned funny' and got jealous of him talking to women in his shop. He packed his bags and left her. He also told me that witchcraft had been dying out until it was formally organized by Gardner. The original witches were the wise women and men of the village – 'real witches were not burnt', he said.

At this point Una came back and expounded at length her views of the origins of humanity. In the meantime the phone rang a couple of times (someone wanted their crystal ball charged) and Lily, an elderly woman whose husband was dying of cancer, called in. Una seemed to run a local surgery of therapy, friendship and support. People milled in and out of the kitchen to make breakfast, coffee, etc. Una told me that the Craft was matriarchal: 'woman is the power and creation', she said. By contrast, 'man is the strength behind the woman – she creates the life force and he protects her'. She insisted that the role of the high priestess was totally feminine: 'to call down the Mother – to be her personification. The Goddess contains maleness, but as perfect essence.' I found this to be an example of the differing, often confusing, views of gender polarity in witchcraft. Una told me that the god form of Pan was the 'personification of all men'. By sympathetic magic the god imitated the power of the animal through shamanism. Una saw the essence of witchcraft in human reproduction, as symbolized by the chalice and the athame. She said:

> Everything has to balance and produce – this is what it's all about – reproduction. Man has tried to deny the female but all of creation will balance. You can't deny the female body – life energy. The denial of the feminine is the denial of existence.
>
> In the proper Craft [a comparison with Dianic witchcraft] all are equal. Man will help woman wash pots, hoover, clean the altar, but woman brings warmth, decorates the altar and brings it alive with femininity. The High Priestess and the High Priest perform the Great Rite to consecrate as one – one mind, one consciousness.

At this point Una broke off to say goodbye to a girl who, she told me afterwards, she had saved from suicide. She was to be allowed to join the coven when she was eighteen.

Wiccan identities are based on sexual complementarity, and do not challenge dominant social gender identities in the ways that feminist witchcraft does. Una thought that feminist witchcraft was mistaken, because it introduced politics into the practice of witchcraft, and this was a taboo area:

You can't bring politics into the Craft – this is what's destroying the Craft. It is non-political. Craft is a family – everything has its place. Religion and politics destroy and bring chaos . . .You can't have all-female [groups] because you need balance, you need yin and yang. Dianic witches are playacting, there is no ultimate balance, so you get 'bitchcraft'. Everything has to balance and produce, this is what it's all about – reproduction.

Una had arranged to go to a Psychic Fair, and she left me to talk to one of her coven members; but before she went she told me about her psychic powers. In 1989 she said she had predicted the fall of Gorbachev to a Russian who had come to her, and she was very upset that she had not been acknowledged as the one to predict the Russian coup. However, she seemed to rely on other psychic powers too. She said that I was OK because the dog had accepted me. He could tell those who practise black magic or have bad vibes, she said. If the dog had growled at me then I would have been shown the door.

As was noted in the last chapter, the wiccan High Priestess Vivianne Crowley uses Jungian psychology as a gendered and harmonizing framework for wicca. She describes the three stages of initiation into wicca using Jungian terminology: the first is an introduction to the Goddess, and involves an opening to the unconscious (to the shadow and the animus and anima); the second, which is similar to the first but includes the *Legend of the Goddess* enacted as a mystery play, is an initiation into the 'mystery of death'; in the third sexual polarity reaches its supreme manifestation in the 'sacred marriage', which according to Vivianne Crowley involves three levels:

> On the physical level it takes place with a priest or priestess; on the psychological level it takes place with the animus or anima; on the metaphysical level it takes place with the Goddess or the God. On the psychological level, the anima and animus can be considered as our initiators into the world of the True Self. Just as it is a priest or priestess of the opposite sex who will perform the rite for us, so too it is our unconscious minds symbolized by the contra-sexual animus or anima who will initiate within us the psychic change which leads to individuation (1989:227).

Here the forces of masculinity and femininity have physical, psychological and metaphysical elements, which may be united if individuation, the harmonious balance of inner wisdom, is achieved.

It was Dion Fortune (1989) who suggested that women, following dominant notions of femininity, were outwardly passive in the ordinary world but dynamic on the inner planes of the otherworld. This means that in wicca men have to give up their social power and concede it to women. The high priestess is the leader of the coven and the high priest her partner. According to Vivianne Crowley, the rationale for this is derived from *The Legend of the Goddess* ritual, when a witch undertakes an initiation ceremony that makes her a high priestess. She enacts the role of the Goddess and undertakes a 'heroic quest', where she confronts the force of death represented as male (1993:135). This stems from Gardner's *The Myth of the Goddess,* which he called 'the central idea of the cult', in which the goddess journeyed to the nether lands to the realm of death: 'Such was her beauty that Death himself knelt and kissed her feet . . .' (1988[1954]). This has been interpreted by the wiccans Janet and Stewart Farrar as: 'Such was her beauty, that Death himself laid down his sword and crown at her feet' (1984:29 quoted in V. Crowley 1993: 135). Crowley notes that in wicca the sword and the crown are seen as symbols of power and legitimate authority, and are given by the God to the Goddess. She says that this 'recognizes the underlying reality of male–female relations: the greater physical strength of the male'. The woman can only rule because the man permits her to do so, and 'In Wicca, he does' (1993:135).

These notions of femininity and masculinity are largely derived from Western cultural stereotypes: masculine typically means outpouring energy, rationality and activity; while femininity usually means containment, intuition and passivity. The aim of traditional magical practices is to balance these forces within the individual in a form of psychological androgyny. The Jungian analyst June Singer sees androgyny as based on the interplay between what are seen as masculine and feminine components of the individual psyche (1989:10–16). Most magical traditions following Fortune's harmony model are based on this same premise. For example, in high magic, which is often very individualistic, the magician learns to balance what are seen to be qualities of femininity and masculinity within her/ himself, while in wicca, femininity and masculinity must also be balanced within, but are also, at one and the same time, fundamental organizing principles of ritual practice, because wiccan ritual is based on the complementarity of gender polarity. In addition, it is traditional practice that a man initiates a woman into the Craft and vice versa.

Thus in wicca gender roles are adhered to – women mediate the goddess, while men mediate the god, and, as was noted in Chapter 4, a central rite is 'Drawing Down the Moon', when the Goddess is invoked into the high priestess by the high priest. I spoke to a female feminist witch trained in a wiccan coven who has drawn down the god into herself. She commented that this was unorthodox, and that she thought that many disapproved. It was difficult for a woman to do, she said, but was possible if she had enough support from the group.[4] The reasons given to me were that this particular group had problems recruiting men. The coven had placed an advertisement in a weekly London events magazine – *Time Out* – which had attracted some very unsuitable men, who it was claimed had dubious reasons for joining a witchcraft coven

Feminist Witchcraft

The introduction, by feminist witchcraft, of sexual politics into the magical sub-culture has created an uneasy relationship with magical practices that hold to traditional gender values, because feminist witches are the only group in the subculture that question the certainties of gender. Feminist witches radically challenge existing social relationships, which they define as patriarchal. They aim to use magic to liberate themselves from those patriarchal social definitions – such as gender stereotypes – that have 'infiltrated' into magical practices. Thus there is an uneasy tension between feminist magical groups and the wider subculture. While wide disagreement is recognized to be undesirable, going as it does against the Pagan ideology of tolerance and diversity, many of the ideas of feminist magic are deeply threatening to other magical practices, and there is often much muted dissent from traditionalists.

A crucial component of the recent history of feminist rebellion started at the Greenham Common peace camp outside the Cruise missile base. It began in 1981 as a political protest by men and women peace campaigners, but evolved into a specifically female protest in 1982. This was a tactical decision based on the idea (unfounded) that the police would be less violent with women, as well as being a more general protest against male domination. The mixed peace camp did not overcome any of the problems of a patriarchal society – the men were more involved in the active organizational tasks, while the women were expected to perform more domestic chores. Many of the women involved moved to a commitment to political lesbianism based on the withdrawal of emotional support to men (Young 1990). From this point onwards there was a constant exploration of a mode of being that was distinctly female, by challenging the assumed 'naturalness' of some gender roles and reinforcing others. Alison Young shows how the protest involved creating a different view of the female body through lesbian relationships and the deploy-ment of female bodies as blockades, which generated a new sense of their power not as sexual objects but as powerful beings (1990:32). This was a time when conventional views of femininity were being disputed and ways of finding a specifically female power were sought through resistance. From my own observa-tions at the time, I would say that the protest drew on essentialist images of women as nurturing mothers in touch with nature and the essence of life. The ability to give birth was contrasted to the male drive for dominance, violence and destruction. The women created symbols and made magical connections. The symbol of the spider as spinner of a fragile, yet strong web became especially important to demonstrate a holistic interlinking of all life, and webs were woven into the perimeter fence to convey opposition to the forces of destruction within the base. There was a cross-fertilization between the women's peace protest movement and the development of feminist magical ideas. Magical practices were seen to offer the space to find or create a self unbounded by patriarchal fetters of feminine gender roles. Patriarchy was associated with dualism, control and division, symbolized by the missile base. Starhawk's writing became central in the search for ideas about re-creating society based on a different source of female power and spirituality, and conventional masculine roles were also challenged (1982:86).

Feminist witchcraft encourages a fluidity – a 'freedom' – from constraining magical gender roles by emphasizing the shamanic or 'shape-shifting' qualities of magical practice. In feminist witchcraft all duality is seen to be the product of patriarchal social construction, and therefore inseparable from power relationships. Starhawk writes that to describe the essential quality of energy flow that sustains the universe as male/female polarity would 'enshrine heterosexual human relationships as the basic pattern of all being, relegating other sorts of attraction and desire to the position of deviant' (1989:9). 'Goddess religion is about the erotic dance of life playing through all of nature and culture' (ibid.). Men in feminist-influenced practices identify with a nature that is fluid, not 'feminine'. There is an emphasis on nature as the animal world: neo-shamanism and 'shape-shifting' – talking to, having sex with, and becoming one with animals in the spirit world. For Wiccans and high magicians 'nature' is feminine, while for feminists it is a genderless other – often animal as opposed to human, i.e. they search for a non-gendered image of otherness.

A feminist contributor to the matriarchalist Pagan magazine *Wood and Water* described her relationship with the Nordic god Loki, whom she called a 'god – against – the – gods'. She writes that when she first became a Pagan she was totally fed up with male deities: 'I'd lived with the Father all my life, and I couldn't see much advantage in the Son either'; but the god Loki had started to 'play a game of hide and seek within her'. Loki is a Trickster, a shaman, and a lover; he is 'Bisexual and shape-shifty, he appears to me to move fluently around the edges and within the crevices of Establishments', and is a force to be identified with as self: '. . . I have been invited to *identify* with these male characteristics that I call Loki. I am not asked either to be his daughter, or to mate with him. I am asked sometimes to participate in his sexuality, but *as myself* (does this make sense – I hope so)' (vol. 2 no. 28. Summer 1989).

The above article clearly shows the exploration of sexuality and identity which a free engagement with the otherworld may bring. The writer says she has been invited to identify with Loki's male characteristics, not as his daughter or as his sexual partner, but to participate in his sexuality as herself. Some Dianic witches often choose not to work with a god, and they are sometimes criticized as 'unbalanced' by wiccans. One writer in the Imbolc issue of *Pagan Dawn* (no. 118, 1996) said that she did not work with a god, and that goddesses had the full range of human potentialities. She would invoke Athene to be assertive, and Kali if she wanted to get in touch with her anger and aggression.

An exploration of the full range of otherworldly potentialities – encounters with otherworldly beings as deities, animals, spirits, etc. – often starts in childhood. Jane, the feminist witch, describes how she became a 'deer person' in her youth:

> I was fascinated by deer, their grace, speed, large eyes and ears, and long slender limbs. I became a deer person, with a basic human body but deer-like eyes and ears, long slim arms and legs, small feet, and a little tail. I covered myself with a thin silky down of hair. I experimented with moving, in a woodland glade, and then, quite unexpectedly, more deer people came out of the trees, and I was part of a group, a little herd.

A group of human-like people from the sea also provided a source of creative visualization for Jane:

> Another area of fascination was the sea, particularly beaches where there were large waves. I would sit and watch them, becoming drawn into the movement and flow, with very intense feelings in my chest, throat and nose. I had (and still have) very intense dreams about the sea, and a group of human-like people came into being, through the dreams, and through watching the sea. They lived underwater, but could walk on the land, and float through the air. I lived this life with very intense sensual and emotional feelings.

Jane also experimented with sexuality in the otherworld:

> Somewhere between the age of eleven and thirteen, I started to experiment with sexual experience on 'the other side'. I started a sexual relationship with my brother, which became very close, intense and powerful for me. I was experiencing orgasms, except that I was too ignorant to understand what was happening. (For all sorts of reasons, I was very late in gaining sexual experience on 'this side', and did not become orgasmic until my mid-twenties. At that point I realised this was a familiar experience from this early phase of my life.) I experimented with being male, with being animal, and created part animal, part human, beings for myself. I kind of invented the idea of homosexuality, before I understood the social reality. I would lie in bed and think about love and relationships, and came to the conclusion that it was people (beings) that were important, not bodies, and therefore it made sense to me that two same gendered people could love each other. I used to wonder about sexual techniques!

These accounts show the fluidity of identity and sexuality that can theoretically be expressed in feminist witchcraft because it does not hold to the gendered polarity of wicca. This magical practice also offers men a space away from stereotypical patterns of sexual identity, and it is particularly attractive to homosexuals or bisexual magicians.

I attended a Beltane weekend workshop run by Starhawk and her husband David Miller on 'Women's Magic and Men's Magic', and organized by Alternatives at

St James's Church, Piccadilly. The structure of the weekend was based on a journey around the five points of the pentacle of life: birth, initiation, ripening, reflection and death. There were around sixty women and eighteen men present. On the Saturday separate women's and men's space was created – the men retired to another room, and the women were left to discover their 'power within'. This is an essential part of feminist magical practice, because women in patriarchal societies have to learn to feel themselves positively rather than through the eyes of men and 'power over'. The self has to be located as the centre of power before that power can be channelled through sexuality.

In the women's session, Starhawk took up what turned out to be her habitual position for drumming and chanting in the centre of the room beside a large altar adorned with participants' personal magical artefacts: candles, figurines, pictures, books and flowers, etc. She started drumming and slowly walking around the altar while she chanted 'I am strong woman, I am story woman, I am soul woman, I will never die.' She guided a meditation trance dance that took the participants though each part of the body from the hair to legs and feet – shaking off what people had said and all the bad feelings about the self. She encouraged the women to be proud of their bodies – to thrust out breasts and pelvises with pride – to get in touch with the power centre – the uterus. Then lying on the floor the women undertook a trance journey to give birth to themselves. Starhawk told a story which those present enacted as she spoke:

> Everyone was around waiting [there was yelling and screaming from the participants]. There were helping hands. Then someone shouted 'It's crowning.' One last push and she's out. Someone shouts, 'It's a girl, it's a girl!' There was great celebration – a girl is reason for rejoicing.
> 'You are lovingly brought out and laid on your mother's belly. You grow up surrounded by people who love you. You can go out and be free.'

The women act out what it is like to grow up free and true to themselves: there was crawling, gurgling, play acting, dancing as they went through each stage of child development. A pillow-fight broke out, and a pillow was thrown around the group. Then Starhawk told about how a girl's first menstruation is celebrated in the Reclaiming Collective. Girls are taken out by the women to exchange experiences about menstruation, and she said that the women were truthful in telling the girls about their periods and what they were like.

The women's magic session ended with the women practising a 'Dance of Power' that was going to be performed in front of the men. The women practised dancing from their power centres, moving in a very rooted and grounded way from their new-found sense of self. After a break, the men returned. I asked one of the men what had happened in the Men's Magic session. At first he was unsure whether it was considered acceptable to speak to me about it, but eventually told me that they had spent the most part of the time discussing how to approach some local boys who kept kicking a football against the outside walls and windows. Further time was spent discussing male solidarity, and the differences between the heterosexual, bisexual and homosexual men. Apparently it was problematic arranging the 'Men's Dance' – finding their own power without being threatening to the women. In the event, it was decided that they should honour the women present first. It was evident that in the context of the workshop the notion of men's power was a lot more problematic than women's power, for obvious reasons. The dilemma that the men encountered had meant that they had to spend more time talking about what they were going to do in their dance, with the result that they were an hour late for the next ritual.

Eventually the women performed their dance, which was strong and assertive. Many picked up coloured scarves from the altar and swirled them around. The men stood around watching. Then it was the men's turn. Their dance started with the men forming a circle around the altar and then facing outwards and bowing down on the ground before the women and honouring them. Then they turned inwards towards the centre, linked arms and chanted very loudly to the green god as they spun around the altar. The women started clapping, the energy dispersed and the men stopped.

In a 'processing' discussion the next day the dances were the subject of much debate concerning issues of power and control. One woman felt that the men had not been paying enough attention to the women's dance and had been talking amongst themselves. Another woman said she felt moved that the men had honoured the women. Yet another woman said that the women's dance had an awesome power that was difficult to hold. Someone else commented that it was difficult in the women's dance to tell if the women were dancing for themselves or for the men, and she felt that the dance contained elements of performance rather than dancing from self. One man spoke about the difficulties of applying the energies of a god because witchcraft was so Goddess – and women-centred. It raised issues of 'How can I be a man in this type of space?' Women could be who they were, but he felt that this was not possible as a man. He thought that it was important to be respectful of women, but that there was a difficult energy between women and men. Another man said that he thought the men's energy manifested in the dance was very male and that it was a good way of getting rid of the stereotypes. Another man said that he had found himself in the male mysteries – he had related to his male self; but that he had subsequently had problems with his female partner who was also at the workshop. He had wanted loving and nurturing, which was unforth-coming from her. Some of the women said that too much time had been spent on talking about the men's dance, and that this was no different from ordinary life – it was what men did that took all the time and energy. This workshop clearly demonstrates the importance of healing to restore lost power and the discussions over the women's dance and the men's dance revealed some of the complexities of 'power within'.

The main ritual on the Saturday evening was the highlight of the weekend: the participants 'visited' the Temple of Desire. During this ritual, which was associated with the 'ripening phase' on the pentacle of life, Starhawk took great pains to establish a Temple of Desire free from any preconceptions about sexual orientation. She created a ritual space that could be interpreted fluidly by heterosexual and homosexual magician alike. Free sexual expression is seen to be a sacred aspect of 'Goddess spirituality', and this runs counter to both high magic and wicca, despite conciliatory gestures from both. Feminist magic thus stands apart from more mainstream magical practice, because feminist witches are critical of magical gender roles, seeing them as oppressive of women.

Starhawk explained that the trance journey to the Temple was both internal and external, to find the place of desire. The Goddess Inanna, was invoked. Starhawk said that Inanna has said that AIDS sufferers are her special priests and priestesses 'whether they have chosen to be on this path or not'. She probably said this because there was a person suffering from AIDS at this workshop, and also because AIDS has affected the lives of many in the San Francisco Bay area, where Reclaiming is based. It was explained that this was a particularly powerful ritual, and because of its power some might not want to visit. For this reason a Pavilion of Healing was set up nearby for those who felt in need of healing energy. Starhawk asked for two volunteers to ground or hold the ritual circle, and for two to help to send the healing to those in the Pavilion of Healing.

The trance journey started with Starhawk drumming and circling the altar. We sang a chant taken from a Mesopotamian myth, the Sacred Marriage of Inanna: 'Barge of Heaven, so well belayed, full of loveliness, like the new moon . . .'. We journeyed to the Temple to find the place of our desire. At the threshold we met a guardian, and those who did not want to pass the threshold could go to the Pavilion[5] (two women and one man went, the rest journeyed on). Starhawk visualized a white, sacred enclosure, which we entered and found within ourselves. After spending some time there – reflecting on our sexuality – the drumming increased and the energy reached a peak and was sent outwards and upwards. A cauldron of flames was actually lit in the centre of the circle, and participants took turns in jumping the flames before retracing steps on the journey back into ordinary consciousness.

In the next section I show how different conceptions of gender influence a form of high magic (Thelema) and feminist witchcraft.

Gender and the Magical Will

The notion of the magical will is central to working magic. Nineteenth-century occultism was largely influenced by the Hermetic tradition of the Renaissance. Much of the forward movement of the Renaissance was due to work inspired by the vigour and emotional impulse of magi who looked back to a pristine golden age of purity and truth. Indeed the historian Frances Yates goes so far as to argue that Renaissance magic may have been a factor in bringing about fundamental changes in the human outlook, which 'may be of absolutely basic importance for the history of thought . . .'. She points out that the Greeks made discoveries in mechanics and other applied sciences, but never crossed the bridge between the theoretical and the practical: 'of going all out to apply knowledge to produce operations'. This, Yates argues, is 'basically a matter of will', because the Greeks valued contemplation above operation. The Renaissance Magus transformed this by changing the will: 'It was now dignified and important for man to operate; it was also religious and not contrary to the will of God that man, the great miracle, should exert his powers' (1991: 155–6). Thus the will became central to magical work. However, the Renaissance magician was male: the male body was taken as the more perfect human model, having certain culturally-ascribed characteristics, and this has important implications for the way that magic is gendered today, particularly in the work of Aleister Crowley. Crowley, drawing on Nietzsche's idea of 'the will to power', implicitly took the male magician as the norm, while at the same time he used women as sexual partners for his sex magick. The magician Kenneth Grant, a follower of Crowley, expresses the notion plainly when he claims that 'The Centre of Will (Thelema) – source of solar-phallic energy – is centred in the priest, while the Fire Snake or elemental Cosmic Power has its seat in the vaginal vibrations of the priestess' (1976:132). Thus it appears that the concept of the magical will and the way it has been developed by Crowley is based on the lines of a specifically masculine model, even though Crowley includes women in principle.

Using the work of Carol Gilligan (1993), who argues that there is a disparity between women's experience and the representation of human development – women do not seem to fit into existing models which take the male as the norm – I argue that the magical will may be seen in different ways by male and female magicians, and that this has important implications for the way that magic is worked. For Gilligan the different experience of women is characterized by Nancy Chodorow's theory (1978) that differential development in boys and girls is due to women's universal responsibility for mothering, which creates asymmetrical relation capacities in girls and boys. Because girls are mothered by someone of the same gender they develop fluid ego boundaries, whereas boys develop their sense of self in opposition to their mothers, and develop more rigid ego boundaries.[6]

Gilligan concludes that relationships experienced by women and men are different: boys experience separation and individuation tied to their gender identity, which is threatened by intimacy, while a girl's identity does not depend on the progress of individuation, but is defined through attachment and is threatened by separation. Women come to define themselves in the context of human relationships, which, when compared to the human developmental model (based on the male experience), are seen as a weakness in comparison with individuation and individual achievement. Much feminist writing has described female identity as 'Other' or as a lack against which masculine identity differentiates itself (for example de Beauvoir 1973).

Magic based on the Crowleyian masculine model relies on solitary practice. This is illustrated by the fact that when Crowley assumed leadership of the Hermetic Order of the Golden Dawn he restored the original rule that members 'should not know each other officially' and 'have as little to do with each other as possible' (Symonds and Grant 1989:562).

Borrowing Rabelais's idea of an *Abbaye de Thélème*, Crowley founded his own Abbey of Thelema in Sicily on the principle of enabling each individual to fulfil their own function. However, this was too much for some: 'We found that life in the abbey with its absolute freedom was too severe a strain on those who were accustomed to depend on others. The responsibility of being truly themselves was too much for them . . .' (ibid.:854). The Book of the Law anticipates this: 'The bulk of humanity, having no true will, will find themselves powerless. It will be for us to rule them wisely. We must secure their happiness and train them for ultimate freedom . . .' (ibid.). Aleister Crowley viewed the development of a magical will as 'solar phallic', and I suggest that it should be seen as a specifically masculine way of working magic. It is Nietzschian in the sense of refering to autonomy, anarchy, control of the universe and explicit power. The new Aeon of Horus involves opening a path to realize the 'True Will', and this path utilizes sexual magick. The magician Kenneth Grant points out that the A∴A∴ (*Argenteum Astrum*) and the O.T.O (*Ordo Templi Orientis*) were two magickal orders that used sexual techniques for 'establishing a gate in space through which the extra-terrestrial or cosmic energies may enter in and manifest on earth' (Grant 1976:136). The 'Scarlet Woman' is the gateway and the means by which this is achieved. Aleister Crowley's attitudes towards women were complex and, according to his student Israel Regardie, were based on his feelings for his mother, ranging from utter contempt to idealization (see Regardie 1986). In addition, Aleister Crowley's model is obviously not immune from dominant notions of gender from the wider society.

In a recent edition of the Thelemic journal *Nuit Isis* Shantidevinath 93 argues that many Thelemites' interpretations of women's role in Thelema are restrictive. She says that that she moved from wicca to Thelema because she saw a new archetype of woman in Aleister Crowley's notion of 'The Scarlet Woman', 'The Lady or Whore of Babalon', which implied liberation and someone fully in control of her own sexuality. However, she argues that attitudes within the practice limit women to one assigned magickal role associated (like wicca) with mythologies of the womb and lunar cycles, connected with the idea that women's power can only be tapped in partnership with a man (*Nuit Isis*, Autumn Equinox 1993 ev). These notions are extremely pervasive in contemporary magical practices.

Gilligan's assessment that women's identity is not based on the progress of individuation but on attachment and relationship is most clearly shown in feminist witchcraft, which may be said to represent the antithesis of Crowley's individualism. Feminist witchcraft, develops what is generally called the 'ethic of connectedness' in its relationship to what is viewed as an evil patriarchy. In so doing the practice provides a clear demonstration of the two respective moral worldviews, based on the individual in the case of high magic and the feminist ethic of connectedness on the other. For feminist witches the emphasis on the true self requires the annihilation of a particularly male ego, and feminist witchcraft covens focus on finding the magical self from an essence of less-bounded ego; they have to strip away layers of patriarchal conditioning to find their true selves. An example from my fieldwork with a feminist witchcraft coven will demonstrate the very different nature of this practice.

It was the time of the Winter Solstice. We purified ourselves in the darkness – each shouted out her anger. One woman was angry with the Goddess because she had fallen for the wrong person again. She was feeling unhappy about the way a relationship had worked out, and this was making her feel bad about herself, she felt there must be something wrong with her. Space was given for everyone to express their negativity. We imagined opening the circle in the dark. The women spoke about how they were imagining the circle. At one moment it was a web, at another a cocoon, spinning around and around past the north over our heads and under our feet. We held hands and spoke about our feelings for the dark and what we wanted from the light. The emphasis was on the changing seasons – the moving from the dark into the light. We then lit the candles of the four quarters. We shifted around to sit where we felt most comfortable, or to where we felt we needed a particular energy. The next part of the ritual was concerned with sending healing. We sent healing energy to an absent member's son who was unwell. He had trouble with his eyes: perhaps he did not want to see the difficulties his parents were having, someone suggested. Then we sent healing to ourselves. Each took a turn in the centre of the circle, while the others shouted and chanted her name as she first stood and then lay on the floor. When she was lying down the chanting changed to a crooning which became an eerie wail. It was quite a strange experience opening my eyes to everyone leaning over me moaning Ssssuuuuuusssssssssssaaaaaaannnnnnnn. When the ritual was over we all shared a meal, which had been contributed by group members, using a common plate.

The emphasis here is on the communal elimination of negativity (due to living in a patriarchal society) and the creation of connection, via web and cocoon imagery, to establish a healing space. The feminist conception of the magical will firstly concerns 'finding' the self from a starting-point of no bounded self. Having found this self it is said the witch's immanent power is realized. Feminist magic is concerned with what Starhawk has termed 'liberation psychology' (1990) – a connection with the earth, egalitarianism and feminist politics. The magical will is realized through the feminist witch's empowerment and the moral resistance to 'power over'. By resistance to patriarchy, the violation of taboos, a reclamation of responsibility for finding internal 'truths', and 'finding safety through solidarity with each other', feminist witches attempt to transform culture and reshape the world (1990:313). The practice is monistic: all is inherently divine in itself once social alienation is overcome – the source of human estrangement is social rather than metaphysical.

Community is seen as a resource for healing divisions; but the practice is more complicated than this, and can become divisive. The practice of magic requires that the individual must come to understand her- or himself, and discover her or his own magical self in a personal experience of the otherworld, while the emphasis on community – as a safe haven from external oppression – prioritizes the magical group. All witchcraft magical groups function as close alternative kinship 'clans', and personal and sexual relationships between coven members run very deep. When people practice magic together they attempt to pull down what they see as barriers of social conditioning constructed by the wider society. Coven members are encouraged to be very open about their feelings and experiences following the feminist view of ideal relationships based on nurturance, caring attachment and mutual relatedness. This may be helpful, or alternatively, it may create a situation leading to the loss of self and the abuse of power. Overall, it encourages an emphasis on an internal community, and this can be antithetical to the practice of magic if the individual is so enmeshed within coven relationships that s/he has lost all sense of her/himself and merges with the group. In this event the whole basis of working magic – the focusing of the forces of the macrocosm in the microcosm and their direction by the magical will – is suppressed.

It has been suggested that the notion of community is problematic for a trans-formative politics in mass urban society (Young 1990; Friedman 1992). The feminist philosopher Iris Young believes that an analysis of the concept of community reveals that the desire for unity or wholeness generates borders, dichotomies and exclusions, and in particular denies differences. Drawing on Derrida's 'logic of identity' (1976), she writes that 'The urge to unity seeks to think everything that is a whole or to describe some ontological region, such as social life, as a whole, a system' (1990:303). This creates mutually exclusive oppositions that structure philosophies – mind/body, culture/nature, male/female, etc. The first member of each pair is elevated, and designates the unified over the chaotic and unformed. This logic seeks to understand the subject as a self-identical unity, but the very act of unity requires expulsion: 'The move to create totality, as the logic of hierarchical opposition shows, creates not one, but two: inside and outside. The identity or essence sought receives its meaning and purity only by its relation to its outside' (Young 1990:304).

The conception of community is invoked by feminist witches to project their alternative to an individualism often viewed in terms of separate self-contained human atoms operating with an ethic of self-interested competitiveness. Starhawk creates such an image of community by contrasting it to 'the culture of death' of patriarchy, whose 'wastes assault our lungs and poison our waters', and whose 'diseases attack our own bodies' defenses, destroying from within' (1990:312). Community comes to represent mutual and reciprocal understanding of persons who relate internally – who recognize themselves in each other because of a shared purpose. The shared feeling of belonging and merging tends to a 'shamanic' ecstatic sense of oneness. In addition, Starhawk suggests that a group is an entity, a 'being with an independent life' (1990:264). When people work magic together they form very deep relationships. These relationships may be seen to be outside patriarchy, but they are not free of power relationships – power is abused within the internal community as well as outside it. There is a widely held belief among witchcraft groups that once a magical circle is purified and opened, and the wider forces are invited to enter there can be no abuse of power within the sacred boundary – yet this is perhaps the greatest and the most dangerous myth of modern magic.

Also the notion of community may bring problems and a denial of difference within a magical group. The political implications of a collapsing of boundaries are that issues of class, race, or sexual difference may be marginalized or left to be resolved magically: for example, the 'class' workshop that I described in the last chapter left differences between classes unresolved on the experiential and intellectual levels. This whole issue raises tensions between the magical identity of the group and the magical development of the individual. Thus it is clear that magical philosophies, such as Aleister Crowley's 'will model' and feminist witchcraft, are shaped by both gender relations and child development.

Sexual Politics

I have shown how magical power is channelled through gender relations and sexuality. The area of gender relations and sexuality has been shaped by the occult frameworks of Aleister Crowley's 'will model', which claims magical autonomy for men and women but is based on a masculine notion of the will, and to a greater extent, Dion Fortune's 'sexual harmony model', which seeks a Jungian complementarity between femininity and masculinity. The introduction of feminism into the subculture has challenged the certainties of gender polarity by opening up an arena for sexual experimentation in relation to magical spirituality; it has also set a stand against what are seen as patriarchal social structures. By contrast, neither high magic nor wicca directly challenge social or political issues, and they thus largely support the gender status quo of female subordination in the ordinary world. In wicca in particular, women's power and legitimate authority is given by men. In Chapter 4 I showed how Gerald Gardner had written that the high priestess rules because the God gives her the power ([1954]1988:31). As Elizabeth Puttick notes, this seems to undermine the basis of female power by defining its conditions as social and as a 'precarious authority granted by men, which can therefore be abrogated by men' (1997:217).

The embodiment of deity and the assumption of the status and power of deity is the basis of magical power in high magic and wicca. Both practices are founded on the notion of spiritual transformation. It is possible that this central element of wiccan practice may derive from Indian Tantra. Samuel notes that the similarity between ancient Indian Tantric and modern wiccan ritual circles is striking. He notes that Vajrayàna or Tantric Buddhism originated in India in the fourth to eighth centuries CE as small initiatory groups who practised rituals in cremation grounds or similar locations. Similar rituals were performed by the followers of the Hindu deity Siva and his consort the Great Goddess, known as Durgà or Kàlì. He claims that wicca is in some ways a modern recreation of something like the ancient Indian Tantra, both having a strong emphasis on ecstatic and trance states and small group rituals where participants undertook behaviour that was unconventional and disapproved of in the wider society (1998:124–5). In Tantric practices women are worshipped as manifestations of goddesses; but, according to Elizabeth Puttick, the main debate is whether the woman in a Tantric relationship is primarily a sex object for the man's enlightenment or whether she is an equal partner. Puttick concludes that such practices do appear to have worked for spiritual enlightenment in the past, but argues that success depends on high discipline and equality (1997: 56–8). It seems reasonable to suggest that there is potential for spiritual transformation based on a genuine equality, but there is also potential for the abuse of power by magicians of either gender. Male wiccans may experience the loss of control in the feminine realm of the ritual circle as sexually stimulating – as an indulgence in the forbidden as a form of eroticism – a subversion of the social world in the tradition of de Sade. The philosopher Timo Airaksinen, in an analysis of de Sade's philosophy, notes that the essence of subversion is to: '. . . undermine all the known rules and priciples, in order to derive pleasure from what is inside, underneath, out of sight . . . The ultimate result is *transcendence* inside, not beyond, the shattered limits' (Airaksinen 1995:2).

Pleasure must be understood as transcendence inside the shattered limits. By contrast, transcendence in magical practices is a crossing of the limits – from this world to the otherworld – that is seen to bring power. It is the breaking of social taboos, which is said to release a great magical power. In this context magical ritual may be a cathartic space of transformation of the social realm; a negotiated space where feminine and masculine gender roles, or any other roles of domination and submission, are enacted as a theatrical play to work out deep-seated psychological issues, which in the process release magical power. In short, magical ritual may be a place where social relationships of power from the ordinary world are negotiated and transgressed. From my own participant observation, and as Bell (1992) has argued (see the last chapter), I suggest that it is difficult to generalize because the power dynamics of any magical group are dependent on the individuals concerned and the particular psycho-social issues that they bring. I started this chapter by stating that, for magicians, invoking the powers of the cosmos is filtered through notions of gender and sexuality – the feminine, as opposed to the masculine is valued; however, only feminist witchcraft offers a practical political model for women's empowerment in the socio-economic world – a vision for transformation of this-worldly reality.

Notes

1. In a talk given to the Pagan Federation annual conference at Fairfield Halls, Croydon in November 1997.
2. There is some ambiguity concerning Ninshubur, David Phelps has pointed out that it is not absolutely unequivocal that Ninshubur is female (personal communication).
3. At a *Talking Stick* Magical Conference held at Conway Hall, Holborn, London on 12 February 1994.
4. It is interesting to draw parallels with the debate about women's ordination in the Church of England. The 'pro-women priests' position claimed that women priests would mediate Christ as humanity, rather than as man, in dispensing the eucharist, while the 'anti-woman priest' argument centred on what was traditional behaviour. While women were capable of dispensing the sacrament, they could not as women mediate the masculinity of Christ.
5. One of the women volunteers spoke of her experience in the Pavilion of Healing later. She said that she had felt that she had been a channel for the energy of the ritual, which had come through her straight into the woman to be healed.
6. I am aware of the Eurocentric assumptions of this theory; however, it remains a useful analytical tool for Western societies.

Magic and Morality: 'A Mixed Spectrum'

On one occasion when I was engaged in a discussion with Pearl, my high magic supervisor, she asked me how I felt about taking responsibility for all the world's evil within myself. This made me think, and I said that I did not feel accountable for all of it. Pearl told me that if I was going to practise magic then I would have to take that responsibility. Magicians see their magical power as coming from the all the forces of the cosmos – both positive and negative. Healing necessarily involves the channelling of forces for good and beneficent purposes, so-called 'white magic'; but the magician must also learn to control what are termed 'negative forces', or evil. In effect, it is impossible to have the one without the other, and in practice magicians often view the forces as a total spectrum. This marks a significant difference between magic – especially esoteric Christian magic – and exoteric Christianity. The former includes what is often called evil, while the latter external-izes it in the form of the devil.

The focus of this chapter, as of the preceding two chapters, is intimately bound up with issues of power, because the magician, when internally 'healed' or 'balanced', is a 'power house' of macrocosmic energies with tremendous potential, and the realization of this capacity is often seen to be a moral duty. Ultimately morality is determined by the individual magician: the source of rectitude lies within. Morality is personal and internal. For magicians the otherworld is the locus of power; but one of the paradoxes of magic is that the otherworld is, at one and the same time, both external and internal to the magician, and a clear contrast is made between internal ethics – which are said to have a connection with otherworldly powers and therefore have spiritual authority – and the social ethics created by ordinary people in the every-day world.

For magicians practising magic means learning to lead one's life guided by otherworldly communication. Such communication from otherworldly beings is experienced and understood through cosmologies and mythologies – which in the Western context have been influenced by Christianity in its esoteric form or as a reaction against it. These communications and cosmologies concerning the other-world inform magicians' actions in this world. The forces of the otherworld are often used to endorse a certain action with divine authority, even though the magician might consider it unethical according to this-worldly norms. In other words, the practice of magic can, in certain circumstances, legitimate what is perceived to be socially unethical or immoral behaviour in the name of higher power.

Central to the whole issue of morality are the different worldviews taken by high magicians and witches (both wiccans and feminists): like other monotheistic religions such as Judaism and Islam, high magic is to some extent dualistic; the notion of humanity's 'Fall' from divine grace and the opposition between good and evil from a basic philosophical and theoretical framework. Dualist worldviews thus oppose the less-than-whole or fallen humanity to the whole which is God. David Parkin has noted that morality is culturally constructed from two basic worldviews: dualistic monotheistic religions, such as Judaeo-Christianity and Islam, and monistic religions such as Hinduism and Buddhism, whereby notions of good and evil are intrinsic to the world. Parkin suggests that morality is associated with a human quest for completeness, and that notions of 'morality' as the means to human perfection and happiness are the 'attempt to avert a kind of cosmic theft'. Thus, he notes, a sense of incompleteness, imperfection or privation can be viewed in two main ways: 'as a necessary or inevitable weakness of a cosmic totality; or as threateningly opposed to the whole of which it was once part' (1985:8–9). In terms of high magic, spiritual evolution – starting with the individual magician as microcosm – ultimately reunites the cosmos. Witchcraft, in contrast to high magic, is monistic, and incorporates notions of good and evil within a totality in a manner similar to Hinduism and Buddhism (Anthony and Robbins 1978). Thus both dualist and monist positions are intermingled in contemporary magical practices and contribute to their complexity and multivocality. Once again, a focus on the otherworld brings order to this complexity: magic may be seen as a human quest for spiritual completeness through symbiosis with the otherworld, and morality plays a large part in reconciling and ordering this process, as I shall demonstrate.

From her research on magicians in London, Luhrmann found that magicians were very concerned about morality, and that they distanced themselves from 'black' magic by stressing that their own practices were 'white': 'They talk about black magic; they usually tell you that there are black magicians elsewhere and stress that they, by contrast, are very white.' She says that inevitably there were unstable individuals who claimed evil powers, but that 'black magic seemed to be a myth, and the talk about it seemed to be part of a general determination to be as morally virtuous as possible' (1986:82). Luhrmann claims that this attitude reinforces two positions: to forbid black magic stresses the increased responsibility of the magician, indicating that magical ritual is effective and that negative sanctions imply real power; secondly, that magicians use ethical constraints to indicate that practice yields consequences. According to this view, moral intention masks the uncertainty of magical efficacy. In asking whether the ritual was efficacious, Lurhmann argues that the intentional satisfaction of the morality may be confused with its instrumental success – whether the set goal was in fact achieved. This technique uses feeling moral to substantiate an uncertain intellectual claim (1989: 89). This functions like a 'mental trick'. She concludes that to magicians the image of the powerful magician is more important than the abstract, pseudo-scientific theory of reality upon which magic supposedly depends, and that magicians do not derive their morals from their rituals, but rather evoke them to give their magic substance: 'through the use of a negative sanction and the assumption of moral intention they deploy morality to bolster the intellectual claims weakened by scepticism and a lack of social support' (1989:91).

While I agree with Luhrmann when she says that notions of black magic and the choice of positive magic over the negative variety are empowering, I disagree with her reduction of morality to its function as a cover for inefficacious 'pseudo-scientific theory', because this argument reduces morality to intellectual claims. Luhrmann's view is essentially sceptical of magic, and does not examine the internal nature of morality. I argue that for magicians the essence of morality comes not from a cover for pseudo-scientific theory but from contact with otherworldly reality, which provides a form of internal knowledge. Joanna Overing has pointed out that an examination of so-called traditional societies has shown that morality is not a domain separate from the pursuit of knowledge and that there are many truths that all have a moral aspect. Overing says the task is to 'understand the knowledge actors have of their moral universe and their standards of validation with respect to it . . .' (1985:5). In the case of Western magicians, otherworldly reality forms the source of moral knowledge.

In this chapter I start with a brief historical overview of magic as part of a moral discourse. This contextualizes the study of the internal source of morality in high magic and witchcraft, and demonstrates the close connection with Christianity through an examination of the battle between good and evil, the emphasis on the light and the power of the dark. In high magic the notions of 'good' and 'evil' are often opposed. In the esoteric Christian school of thought associated with Dion Fortune the energy or power of evil is seen to promote the good in the sense of spiritual transformation to the 'light' of spiritual awareness and union with divinity. Another tradition in high magic stems from the writings of Aleister Crowley and focuses on the 'dark' as opposed to the light in a similar search for equilibrium and 'gnosis'. I then move on to look at witchcraft. Notions of morality in witchcraft are influenced both by Aleister Crowley's rebellion against Christianity and also by the romantic anti-Christian tradition stemming from such nineteenth-century writers as Jules Michelet and Charles Leland. Morality in contemporary witchcraft is focused on the Wiccan Rede 'An it harm none, do as you will.' In comparison with high magic, the Wiccan Rede is a rather vague moral maxim to follow. It repudiates dogma in the name of individual freedom; but this often means that it becomes dogmatic in its anti-dogma, resulting in the fact that it is difficult for witches to have any sense of communal ethics. The complexities it engenders as a practical ethics are discussed. A comparison of morality in high magic and witchcraft completes the chapter.

Moral Discourse: Magic and Demons

I start this section by looking at how concepts of magic and witchcraft developed in Europe as part of a moral discourse. Anthropologists used the term 'witch' to refer to indigenous ideas regarding anti-social behaviour. For example, Evans-Pritchard (1985[1937]) demonstrated that Azande witchcraft functioned as a natural philosophy to express what was considered socially good or bad through the dynamic interrelationships of people in inauspicious circumstances. Mary Douglas has commented that the symbols of witchcraft are built on the theme of vulnerable internal goodness attacked by an external evil power, and argues that witchcraft beliefs, the beliefs of ordinary people about agents of evil, are essentially a means of clarifying and affirming social behaviour and social definitions: in short they are a communication system (1970). Norman Cohn, looking at what he terms 'Europe's inner demons', goes further, and calls the fears of external threats to community a fantasy whereby another society, small and clandestine, threatened the existence of the larger society by its addiction to abominable anti-human practices (1993[1975]:iv). Cohn charts this fantasy from the second century, when Greek and Roman pagans accused small Christian communities of ritually sacrificing children. This eventually contributed to the construction of the stereotype of the witch and the subsequent European witch-hunt. This stereotype came into being as a result of the Inquisition's campaign against Catharism in southern France and northern Italy. The execution of the first living example of the stereotype was conducted in 1275 in Toulouse. The first mass trial and execution of witches was also held in Toulouse in 1335. However, the fully developed stereotype – of a person who flew by night to attend the sabbath – was first officially sanctioned by the Church not in the early fourteenth century but a full century later (1993:203). Cohn traces the development of the European/Christian stereotype of the witch from ritual magic. He argues that the great witch-hunt was concerned with ritual magic, and that 'over a period of generations, ritual magic and the struggle against ritual magic helped to produce the fantastic stereotype of the witch' (1993:102).

In mainstream discourse 'magic' has been associated with the maliciously manipulative, and connotes the dark and the alone. This usage has a long tradition. The historian Valerie Flint explains how the practice of the art of magic has included vanishings, changes of shape, stature, and sex, transformations into other creatures, night visions and flyings, raising storms, scattering thunderbolts, transporting crops and cattle, control of love, abortions, injury and death. Increasingly magic was represented by the word *maleficium*. Magic during this period was viewed as real and threatening: it was always potentially evil, and might become uncontainable.

She notes that in the Bible too, there was condemnation of magical practices. The Old Testament God was averse to divination (Num. 22:7, 23:23), to augury and to necromancy (Deut. 18:10; Samuel 28:7–25) to mediums and to wizards (Lev. 19:31, 20:6,27), and to all forms of enchantment and shape-shifting (Exod. 7:10–12) (1993:15–18).

Flint argues that in the early Middle Ages an 'enormous variety of religious cults' – Zoroastrian, Neoplatonic, Jewish, Gnostic, Christian – were linked together by a belief in demons, as spirits of evil were held to be real and powerful agents of human misfortune and the possessors of supernatural powers that flew or floated about, awaiting their many opportunities to cause plagues, famines, tempests, stormy seas, sickness, and deaths (1993:18). Humans could control demonic malice in one of two ways: by subjugating them by their own sanctity, or by co-operating with them by invocation. Flint concludes that by the early Middle Ages there was a stark opposition between sanctity and magic, and that magic as *maleficium* carried a 'heavy weight of condemnation'. However, she notes a countercurrent that rescued certain practices and created a distinction between good and bad magic: a distinction between arts aimed at the injury of humans and predictive measures aimed at protecting crops or assisting in cures. The distinction was thus made between magic that was bad and magic that, 'under another name and in different hands', might be good: 'The operations of a little science and much religion might now allow a good deal of magic to slip across the borders', and magic was thus given a passage through the Middle Ages on a Christian vehicle, albeit under another name (1993:31). Nonetheless, the magician, one who specialized in magic, was still associated with *maleficium*.

During the Renaissance the practice of magic was re-interpreted in the light of Hermes Trismegistus and the Kabbalah, and the bringing down of what were seen as higher angelic forces. This had the effect of changing the status of the magician from evildoer to agent of God and the good. According to Yates, there was a change in the status of the magician. The necromancer and the conjuror were both outcasts from society, regarded as dangers to religion and very different from the philo-sophical and pious Magi of the Renaissance. Yet Yates argues that there is a kind of continuity between them, because the techniques are based on the same princi-ples: 'Ficino's magic is an infinitely refined and reformed version of pneumatic necromancy, while Pico's practical Cabala (Kabbalah) is an intensely religious and mystical version of conjuring' (1991:107–8). Yates notes that Ficino was 'too pious and careful' to use planetary or zodiacal demons. He was content with natural magic, which was 'harmless but weak' (1991:91). By contrast the magus, who combined natural magic with Kabbalah, was in a different position. Angels or divine spirits and bad angels or demons are a part of the Kabbalistic system, which is a way of both capturing the powers of the superior higher forces and at the same time of using the powers of the evil. This is the tradition that modern magicians have inherited, and I now turn to an examination of contemporary magical practices, noting the difference between Christian and pagan interpretations.

The Christian tradition of magic primarily stemmed from Pico della Mirandola and the incorporation of Kabbalah, while the pagan strand had its origins principally in Giordano Bruno's attempt to take Renaissance magic back to its Egyptian pagan source. Thus in high magic morality has been shaped by its Judaeo-Christian roots since the Renaissance and Pico della Mirandola's combination of magic and Kabbalah. Western high magic has taken the religion of the dominant culture and interpreted its practices as a battle between good and evil forces within an evolutionary cosmology. This is the tradition inherited by the Hermetic Order of the Golden Dawn, the emphasis being on spiritual improvement and cosmic evolution – on overcoming 'the Fall'. However, Giordano Bruno's emphasis on the pagan sources has been an underlying current, and one on which Aleister Crowley has drawn to fuel his anti-Christian rebellion, with his emphasis on the 'dark' and the anti-social. His aim was also evolutionary, however: not in the sense of spiritual improvement and the 'Light', but rather in terms of Nietzsche's will to power, individual autonomy and anarchy.

Witchcraft (both wicca and feminist) shares the anti-Christian perspective of Aleister Crowley; but witchcraft is a communal practice, and this is inherently problematic. Morality in witchcraft is vague – the Wiccan Rede 'An it harm none, do as you will' supports individualism, but is the source of much debate in the subculture as to how it can form the basis of a practical pagan ethic. By contrast, feminist witchcraft associates all evil with patriarchy: once released from the current capitalistic patriarchal system of domination – what Starhawk terms 'power over' – all humans are in touch with their 'true selves' or the 'power within', and will by definition act in a moral way.

The different aspects of morality come together to form a complex, contradictory, and often confusing array of different ideas and values, and I must also emphasize again that practitioners themselves often move freely between practices, either consecutively or at the same time. The moral dimensions of magic are based on the notion of finding the 'true self', and this encompasses a sense of internal coherence or 'morality' in harmony with the cosmic forces.

The Internal Source of Morality (1): High Magic

The Judgement of Osiris

In high magic morality starts within the self. At a high magic ancient Egyptian ritual drama that I attended, called 'The Judgement of Osiris', the high-ranking magician who took the part of Osiris was to be judged as a moral lesson to those present. The judgement of Osiris was significant because, as first king of Egypt, he was said to have established justice on the banks of the Nile (Baring and Cashford 1993:228), and thus represented order in the Egyptian pantheon. The death and resurrection of Osiris is the prototype of spiritual regeneration, and demonstrates the practical ritual application of notions of good and evil. Seth (Set), the brother of Osiris, is represented as an evil force opposing spiritual progress; yet he is an essential element of it, because his act of murder causes the resurrection of Osiris: the evil fuels the good (Cavendish 1975:176).

The ritual drama dealt with the forces of good, as represented by Osiris, and the forces of evil, represented by Seth, who symbolized the unregulated chaotic powers of the universe. We were told that Seth had been 'conceived with a flaw so that humanity would have the experience of evil, thereby to know how to achieve perfection'. Seth had murdered Osiris 'And the Gods [Geb, Shu, Hapi and Ma'at bearing the body of Osiris, Atum-Ra, Isis Nephthys, Nuit, Hathor, Bast and Thoth] are gathering in the Halls of Judgement'. Osiris's body is placed upright in a sarcophagus set between the Pillars of Jachin and Boaz (which represent all polarity): 'It is foretold that Osiris will live again and he will become as a Judge to those who pass over into Amenti in the West. Before the Lord of the World will come all men and women, each in their time, and he will judge their deeds in the upper world.' But first Osiris himself must be judged. Osiris is judged by forty-two assessors, chosen from the audience, who proceed to question him. Everyone was given a printed copy of the questions so that they could follow them. The part of Osiris was played by a senior magician, who, it was explained, was answering the questions truthfully, thoughtfully and honestly from his own heart:

Assessor 1: Hast thou cared for thy body, as a gift of great worth?
Assessor 2: Hast thou lived the full time allotted thee, and not turned thy hand against thy self?
Assessor 3: Hast thou been clean in mind and body?
Assessor 4: Hast thou loved the body only where the heart is also?
Assessor 5: Hast thou had knowledge of those forbidden to thee?
Assessor 6: Hast thou kept only to the sword, or to the distaff?
Assessor 7: Hast thou respected the youngest brethren of the earth?
Assessor 8: Hast thou stolen?
Assessor 9: Hast thou taken food and drink to excess?
Assessor 10: Hast thou killed?
Assessor 11: Hast thou spoken unjustly of a fellow man?
Assessor 12: Hast thou looked upon the goods of others with envy?
Assessor 13: Hast thou been jealous of another's wife?
Assessor 14: Hast thou told lies of any one in anger?
Assessor 15: Hast thou been lazy?
Assessor 16: Hast thou profaned the mysteries?
Assessor 17: Hast thou given in to false pride?

Assessor 18: Hast thou strayed from the path of thy life?

Assessor 19: Hast thou lusted for precious metals?

Assessor 20: Hast thou been too worldly?

Assessor 21: Hast thou been just in thy dealings in the market place?

Assessor 22: Hast thou repaid debts as promptly as possible?

Assessor 23: Hast thou been generous to the needy?

Assessor 24: Hast thou used lies to gain from others?

Assessor 25: Hast thou used laughter as a weapon against others?

Assessor 26: Hast thou been a friend?

Assessor 27: Hast thou hated another to the exclusion of all else?

Assessor 28: Hast thou lent thy body to any from the other side of life?

Assessor 29: Hast thou been a joy to thy parents?

Assessor 30: Hast thou honoured all faiths that are of the Light?

Assessor 31: Hast thou taken time to be at peace with the Gods?

Assessor 32: Hast thou rejected advice given in love?

Assessor 33: Hast thou listened to that which was not for thine ears?

Assessor 34: Hast thou lived in the Light?

Assessor 35: Hast thou been a sword for those weaker than thee?

Assessor 36: Hast thou enslaved another life?

Assessor 37: Hast thou faced the mirror of self?

Assessor 38: Hast thou taken the words of anyone as thine own?

Assessor 39: Hast thou understood that life ends only to begin?

Assessor 40: Hast thou remembered thy brethren of the earth and been compassionate to those that serve thee as beasts in the field and in the home?

Assessor 41: Hast thou ever worked man or beast beyond its strength?

Assessor 42: I am the last Assessor, but mine is the most important question of all. If the answer be yes, then Amenti is before thee. Is there one who is glad that thou hast lived?

Group 1: Bright is the day that shall come, for Osiris lives in Amenti!

Group 2: Let humanity rejoice, for Balance is come into the world!

After the ritual we were told that we must absorb and answer the questions ourselves, as we too would also be judged in our turn.

Connection with the Spirit and the Higher Self

Much high magic philosophy concerns a coming to understanding of the ordinary self in an attempt to reach the higher self (the magical self). At a high magic conference in central London a High Magus, who had been flown over from the USA, led a ritual on the second day. She had spoken the previous day on why, as a black woman, she had not got involved with Yoruba or Voodoo.[1] The 'inner beings' had told her to do high magic. 'We are all the same on the inner planes. When I connected with the spirit I knew who I was', she said. She had travelled from the USA with her entourage of six female magicians and one young lad of about thirteen. On the following day they were involved in an emotional ritual that had strong moralistic overtones, and that a number of the participants found disturbing.

After the usual robing in the ante-room, some four hundred robed figures filed into the room prepared for the ritual while music played. Another High Magus explained that we were going to be working only with white light. He cast a circle with the sword and purified us saying that we would only be working with the forces of good, and with his sword invoked the four guardian angels – Raphael, Michael, Gabriel and Oriel. The group from the USA came in wearing elaborate white linen robes and sat at the altar in the East, which was decorated with white lilies and irises and lit with candles. Then the American group performed a ritual dance. Four of the women had large blue stars sewn on the back of their robes, while the two oldest had golden circles with Hebrew letters on the back. The young lad, wearing a simple white robe, stood in the middle holding a staff topped with antelope horns.

When the dance had finished, the magus from the USA got into a trance state and began a long and difficult 'sermon'. She asked that everyone under fifty years of age should sit down. Only a few were left standing. She asked that we who were sitting should respect the elders – 'The elders know', she said. Then she started a long tirade about how we thought we were teachers, yet we knew nothing. The magus seemed to sink deeper and deeper into this. 'We were not worthy, and anyone who was not worthy had no right to be part of the Mystery School. We had to earn the right to be part of the School, even if this meant that the School was only left with one person.'

Her voice got lower and lower and it seemed as though she was channelling something dark and terrible. Then she toppled off her chair. All the high-ranking magicians rushed around her as she started moaning. I was not sure if she was having a heart attack. Her fellow American magicians surrounded her and started to sing to raise her spirit. After what seemed an eternity she was eventually carried out, supported by all the magi. The magician at the door, who was guarding it with an enormous sword, let the party out. Previously he had been reluctant to let anyone out – many had tried to leave, as they felt uncomfortable with the proceedings; one woman had almost screamed in her anxiety to go. The magus was taken outside and revived with drinks and lots of stroking and talking, while the congregation inside were told to clap and beat the floor with their magical staffs to rally her spirit. After a long period of time she returned amid clapping and general hysteria, and went around the room hugging people. One old woman was weeping uncontrollably. The American magus, looking rather weak, proceeded to initiate two other magi to adepthood. She anointed their heads while they knelt in front of her. Then she drew down their magical names. She led them around the circle calling out their new magical names as: xxx son of xxx (his real mother) seven times, while we clapped.

Next we recited the 23rd psalm, but starting 'The Goddess is my shepherd, I shall not want . . .'. Then the American Magus did a healing meditation for us. She called down purple healing light to flow through us, then emerald green, then white; all negativity was to be put in Ain Soph, the 'unmanifest', a plane of 'Negative Existence' that 'transcends thought' (Fortune 1987b:32, 33). She called for any who wanted healing (for themselves or for others) to come forward and kneel in front of the altar. This was a very emotional moment, and everyone was crying and sobbing. The two new adepts came and touched everyone. The Lord's Prayer was said. Then it was over, and every-one filed out looking very drained.

This was a very disturbing ritual for many people. When I went outside the room where the ritual had been held, I noticed that there were small groups of people sitting huddled in corners weeping and discussing the emotions that had been raised for them. I too felt very unsettled by this experience, and was unsure about what had happened. I spoke to one woman who had been very critical during the ritual, and she told me that she thought that the magus could not control the forces she was mediating. When I sent my magical diary record off to my magical supervisor for comments, I asked her about the ritual. She replied in her typically contradictory way that the magus was not directly criticizing those present, but was talking about our higher selves and the Inner Teacher of the school. My magical supervisor explained that trancework could have some strange effects, which were sometimes upsetting, and the next time I would realize that unforeseen things might happen. She said that the magus' spiritual contact was 'something quite ancient' – a winged lion associated with the South – and something 'punchy and dynamic'. She explained that the level of energy in the ritual had dropped because the magus had taken it to mediate, and that mediation passes through the subconscious and the subconscious may divert the energy (sometimes by 'only a half-heard criti- cism'). While she acknowledged that a number of magicians had criticized the ritual, she played down that fact, and said that I should look at it as interesting. She said that it was the one time in my life that I would see the Coptic tradition, which was 'all kinds of things mixed in – a bit of Catholicism, ancient Egyptian, Voodoo, and not something you are going to find at every street corner'.

My overall view of the ritual was that it was originally planned to be one with a moralistic message addressed to our higher selves. The magus had pinpointed the discrepancy between our present state and the ideal aspiration associated with the cosmic forces. It reminded me of going to Sunday School as a child, and the moral emphasis on how good one ought to be. As an anthropologist, I was faced with the deep emotion and near hysteria that had been generated by this large group of high magicians. It was impossible to feel detached. This ritual demon- strated the power of magical ideologies precisely because they were experienced directly through the body. They demonstrated Durkheim's emphasis on the social nature of religion as a force, and on its powerful impact on the individual. Magical rituals structure magicians' worldviews and give meaning to their practices in a very direct, deep and internal way. For the magicians present, the ritual was a clear demonstration of the power of the forces of the cosmos. It was conducted within an unorthodox esoteric Christian conceptual framework: the emphasis was on the light, and a battle between good and evil forces was conducted within the body of the magus during a trance; it was a graphic example to those present that 'evil forces' really do exist, and took the battle between the forces of good and evil beyond the level of philosophical debate.

It thus becomes clear that a high magician has to balance and use the energies of good and evil within the self. Dion Fortune stated that the biggest difference between orthodox exoteric Christianity and magic is that the former ignores the rhythm of good and evil; that is, orthodox Christianity is antagonistic and static, whereas magic is equilibrating and dialectic (1987a).

The Battle Between Good and Evil

The division of cosmic forces into positive and negative or good and evil is highly developed within high magic philosophies, but the Christian focus on the 'light' has largely dominated the practice of magic from the end of the eighteenth century to the twentieth century, largely through the influence of Rosicrucianism and the Hermetic Order of the Golden Dawn. Such a Christian interpretation of magic is more in sympathy with the dominant culture in the wider society. To all high magicians, life is polarized between divine and demonic forces of energy, but for those who emphasize the light, evil is seen as a very necessary force whose energy is converted into the power of good by the magician in his or her attempt to find the true self in union with the ultimate pure spirit. This view stems from the alleged fifteenth-century writing of 'Abraham the Jew' to his son Lamech, originally written in Hebrew. A seventeenth- or eighteenth-century French translation, held in the Bibliothèque de l'Arsenal, was in turn translated by one of the co-founders of the Hermetic Order of the Golden Dawn – Samuel Liddell MacGregor-Mathers.

In the introduction to his translation Mathers writes that the author of the work, Abraham the Jew, appears to have been born in 1362 and written the manuscript in 1458 when ninety-six years old. This would make him a contemporary of Christian Rosenkreutz, the founder of the Rosicrucian Order.[2] Apparently, like Rosenkreutz, he had a desire for magical knowledge and travelled in search of 'Initiated Wisdom'. The culmination of his wanderings was his meeting with 'Abra-Melin[3] the Egyptian Mage', from whom he received the teachings that form the second and third books in the original manuscript. The basis of the system of secret magic of Abra-Melin the Mage, as taught by his disciple Abraham the Jew, concerns the invocation of Divine and Angelic Forces to control Demons: 'by purity and self-denial to obtain the knowledge of and conversation with one's

Guardian Angel, so that thereby and thereafter we may obtain the right of using the Evil Spirits for our servants in all material matters' (1976:xxvi). This ritual is similar to that described by Ronald Inden of the Pancaratra Vaishnavas, a major order of Hinduism, which involves a confrontation of higher and lower selves. The Vaishnavas conceive evil – 'papa' or 'tamas' – as opposed to 'punya' or 'sattva', which is goodness, and the work of good over evil is conceptualized by three strands: upper, middle and lower. The first strand is the highest and most enlightened, and is associated with the intellect. By contrast, the third strand represents lethargy and degeneracy, and is the source of evil, and associated with the senses. It is, however, conceived of as an integral part of the cosmos. In between is the second strand, which is passionate energy. Humans need to do good, to uphold Dharma and the cosmological order, by using the passionate energy of the second strand, to fuse with the higher substratum of enlightened existence to stand against the lower substratum and transmute it into 'triumphant glory' (1985:151).

Mathers defines magic as 'the Science of the Control of the Secret Forces of Nature', and he writes that the invocation of 'Angelic Forces' and 'Ceremonies of Pact' with and submission to 'Evil Spirits' form the basis of this type of magic (1976:xxv). However, the magic of Abra-Melin the Mage takes this conception further, in a practical magical application that gives direction for the 'Good Angelic Forces' to control the 'Evil Spirits'. The 'Good Spirits and Angelic Powers of Light' are superior in power to 'Fallen Spirits of Darkness', who are punished and condemned to the service of 'Initiates of the Magic of Light'. As a consequence, all ordinary material effects and phenomena are produced by the labour of Evil Spirits under the command of the Good; but whenever the Evil Demons can escape from the control of the Good they will try to make man their servant by inducing him to make Pacts and Agreements and by trying to obsess him. The Adept, therefore, has to dominate them by his will, purity of soul and intent and power of self-control, which is only attained by self-abnegation. Man is controller of the middle realm between angels and Demons, and has attached to him a Guardian Angel and a Malevolent Demon, as well as spirit familiars. In order to 'control and make service of the Lower and Evil, the knowledge of the Higher and Good is requisite . . .'. The general thrust of this work is 'by purity and self-denial to obtain the knowledge of and conversation with one's Guardian Angel, so that thereby and thereafter we may obtain the right of using the Evil Spirits for our servants in all material matters' (1976:xxvi). These magical ideas of Abra-Melin the Mage have had a profound influence on contemporary high magic, and are the basis for the Abra-Melin ritual as described in Chapter 2.

The Emphasis on the Light

For contemporary high magicians the internal balancing of forces is viewed as a battle between the higher and lower selves – of celestial forces against the ego, which stands in the way of spiritual realization. Thus the invocation of celestial influences implies the expulsion of the lower, bestial or demonic creatures that ordinarily inhabit the mind – the demons of desire and hatred. According to the magician Arthur Versluis these demons are the 'personifications of the forces within each of us which govern our everyday lives. Each time we manifest desire or aversion, we are bringing them to life and signing a pact with one of the demons of ego' (Versluis 1986:60,61). The celestial forces of good and God must overcome the evil lower self and ignorance. The theory behind this is that the individual is made up of an upper, middle and lower self. The energies of the lower self must be utilized in the transformation to the upper higher self. This is the aim of the Abra-Melin ritual. William Gray, a high magician and one-time member of Dion Fortune's Fraternity of the Inner Light, explains that the Abra-Melin ritual is a 'confrontation between . . . a conscious dedication of one's Devil self to one's Deity self' (1989:99). During this ritual the magician must examine his/her conscience and character deficiencies. These deficiencies are equated with specific demons, and are called into consciousness. The aim is a sort of therapeutic spiritual alchemy, transmuting the lower to the higher. It is this higher self that becomes personalized and forms the basis of the magical identity:

> I am not talking to the human side of me, but to that Greater Being of Whom I am but a minimal fraction. I am trying to make a better human creature out of myself so that your Infinite Identity will also be improved to that extent. Therefore I am attempting here to become more clearly conscious of my superior spiritual self which can come into closest contact with you. So I am asking you to increase my awareness not only of that selfstate, but most especially with it, so that I shall know for certain what you mean inside my mind and with my limited intelligence. Let me feel and experience the actuality of my connection with you, so that this becomes as an advisory Angel guiding me in the golden way of goodness and guarding me against all adversaries and Satanically set snares. I will wait with patience while you work with Power inside my individuality (1989:121).

High magicians utilize the terms Satan and the Devil to symbolize negative traits in the psychodrama of the Abra-Melin ritual in a confrontation between what are seen to be the good and evil parts of the personality. If the magician is able to balance the two forces they will combine and transmute to the Divine Being. This is done by direct confrontation on their own ground with devils, who are commanded by the magician. Sins or failings are classified into divisions under the heading of some specified fiend, and are then called into consciousness. Gray describes it as a process of individual evolution, of becoming one with the Divine Being. This is realized by using energy from the Devil: by employing those forces in human nature that have been called evil and that alienate the magician from Divinity and enslave his or her soul for their own purposes; this makes the magician into 'fuel for the Devil's dynamos', thereby providing him with power for working wickedness. The aim is to convert evil power for the benefit of good (1989:108).

In this account the Devil has an important role to play in ultimately returning humanity to oneness with Deity. However, Gray warns of the dangers of the development of a 'pseudo self'. If the energies of good are directed towards the principle of evil then the pseudo self can develop into an alternative being which is the antithesis of good, in short, the Devil. The Devil stands for chaos, and is the antithesis of everything that is ordered, regular, beneficial. Thus the power that comes from oneness with the Deity is dangerous because the energy that gives power to the good is also the power that destroys and will turn to evil if it is misused or unbound. In this school of thought evil is viewed as a 'thrusting block for good'. Dion Fortune explains that there is a type of evil that is a force that can be used for the creation of more good, and there is a type that must be destroyed (1976:14). In this magical view angels and demons are seen to be within the self. A demon may be a stray thought interfering with concentration, a 'complex', an 'unsatis- factory trait', such as laziness, or a driving-force such as hatred or cruelty. By comparison, the guardian angel is said to be experienced as the 'true self' (Cavendish 1975:254).

However, Richard Cavendish, the historian of magic, notes that for practical purposes it is more convenient to assume that spirits, both good and evil, exist independently of humans (1975:255). The magician is presumed to be a person highly trained and experienced in contacting and channelling the forces of the otherworld. This may be for good and beneficent purposes, as can be seen from an examination of how Dion Fortune channelled forces for good against Hitler during the Second World War. High magical practices have been utilized to defend what is seen by magicians as the greater good. Seeing the 'Group Soul of England' under threat before the Second World War, and working on the magical assumption that 'circumstances reflect consciousness', Dion Fortune organized a magical meditation group to invoke the cosmic forces of good to strengthen it. These letters, together with a commentary by Gareth Knight, a leading magician from the School, have recently been published by the Society of the Inner Light (1993). Fortune wrote weekly letters to her students between 1939 and 1942 to confront the evils of Hitler's Germany, the aim of the letters being to maintain the strength of the group mind of the nation against attack. Every Sunday morning from 12.15 until 12.30 members were invited to participate in a group meditation. Some members meditated at the London temple, while others were 'scattered all over the country' (1993:1). The aim was the reform of Germany and the dispersal of its dark psychic heritage by the Power of Christ. This could be achieved by visualizing Germany as a 'vast marsh of human helplessness and ignorance covered by the thick black cloud of ancient evil', through which the sunlight of the Christ could not reach. Angels of the Lord were imagined hovering over the cloud of ancient evil and directing lightning flashes from the points of their swords so that 'the Sunshine of Christ shines throughout the land' (1993:48).

This is a very explicit example of magical work using visualizations of light to work against what are termed dark forces or 'evil energies'.

The issue of how to approach evil forces is still very current in high magic, and is frequently discussed. At a conference that I attended a well-known magician spoke of the 'sinister side' of magical practice. He said that Hitler was initiated into the Thule magical group: 'Hitler's spirit visited planes that he shouldn't have visited – he was made to believe that he was the new Messiah and he made a blood sacrifice of his own people to the Teutonic gods of war.' This 'dark side' is, according to this speaker, 'as bad as it gets'. He believed that after the defeat of the Nazis by military means, the group mind of Nazism still exists on the inner planes and that it will always re-emerge. He spoke of how 'the distortion of esoteric truths has always been with us', and proceeded to explain how Dion Fortune became a magician to defend herself from such abuse and psychic attack, and how Aleister Crowley had become addicted to heroin, and his magic had become so distorted and twisted that he 'had a history of doing funny things to goats'. He intimated that many magicians were not able to cope with their own lives and emotions, asking why the pattern of a magician's power goes up and up and then drops. The speaker told of how he himself had had two full-scale nervous breakdowns due to his esoteric training, which stirs up the unconscious mind: 'the dark side is always with you, and the danger of ego inflation'; the answer is to 'Know Thyself'. For this speaker there was no fundamental evil, but only good that was twisted: 'a person is incapable of evil when the distortions are stripped away'.

However, much high magical practice concerns the balancing of polarized energies. The Kabbalistic Tree of Life glyph is arranged on two pillars: these are the symbols of polarity. A candidate for initiation into high magic stands between these two pillars, which are said to symbolize the candidate's balance between the two. The candidate thus makes a third pillar – the Pillar of Equilibrium. In magical philosophy all things have a dual aspect: positive–negative, volatile–fixed, mercy–severity, past–future, progress–reaction, all of which are contained within each other and are needed for the harmonious functioning of the microcosm and macrocosm.

Some high magic, in its emphasis on the good, does resemble more orthodox exoteric Christianity, and some other magical practices have reacted to this view and focus on dark chaotic forces: an example are those who work on the 'Qliphoth', to which I turn.

The Power of the Dark

It is said that when cosmic forces are invoked, they always come up in pairs – as action and reaction. Every force has its unbalanced or Qliphotic element. The Qliphoth are said to be inharmonious forces that came into existence before equilibrium was established, and have been 'reinforced by the mass of evil thoughts ever since' (Fortune 1987a:50). The ten spheres of the Qliphoth represent the negative aspect of the ten spheres on the Kabbalistic Tree of Life, and Fortune considers them to be 'ten sinks of iniquity' filled with evil 'since the days of Atlantean Magic, through the decadence of Babylon and Rome, down to the Great War'. They are formed from the workings of 'Black Magic', which then assumed an independent existence, and now appear as dreams and hallucinations, and may produce objective phenomena, such as 'noise, deposits of slime or blood, balls of light, and, above all, stenches of an amazing pungency' (1988:90). She warns that those who enter the 'dark portals which lead to the dread subterranean palaces of Qliphoth' return from this journey with their 'bias towards evil intensified' (1987a:112).

Fortune's emphasis on Christianity and the morality of the Light stands in opposition to Aleister Crowley's anti-Christian views, which are also well represented among contemporary magicians. Some magicians claim that too much attention is paid to the light in working with the Kabbalistic Tree of Life, and that Western occultism has been dominated by interpretations that take into account only the positive aspect of this glyph. They seek to develop Aleister Crowley's system of magical working, which, as has already been noted, stems from Bruno's pagan magic, and, following his teaching, place great emphasis on incorporating sexuality.

Kenneth Grant, a follower of Aleister Crowley, argues that the negative or averse side of the Tree has been ignored, and this creates imbalance, because 'there is no day without night, and Being itself cannot be without reference to Non-Being of which it is the inevitable manifestation' (1994[1977]:1). Grant defines the 'infernal realm of the Qliphoth' as a world of shells or shadows – 'the world as we know it but without the transforming light of magical consciousness'. The Qliphoth is the plural form of *qlipha*, meaning 'a harlot' or 'strange woman', which he says are terms that signify 'otherness' (1994:275). The Qliphoth is 'a dark web or nocturnal network of paths, the very existence of which is denied or ignored by those who are unable to realize the total truth of the Tree and its validity for those who would climb even its lower branches' (1994:2). It has been associated with evil because the forces of 'Non-Being' have not been properly understood, and have been cast as powers of 'evil'. In consequence, myth and legend are 'alive with demons, monsters, vampires, incubi, succubi, and a host of malignant entities all of which are symbols that conceal unnameable, unknowable concepts of Nothingness, Inner

Space, Anti-Matter, and the ultimate horror of Absolute Absence' (1994:142). Grant sees the sephiroth as globes rather than spheres, and the paths between as tunnels 'boring deeply into space' rather than flat. Entry to the Qliphoth is gained via Daäth, a 'black hole' or gateway to the parallel universe represented by the otherside of the Tree of Life. 'Daäth is knowledge of the phenomenal world reflected downward through the Tunnels of Set at the back of the Tree' (1994:25).

Some magicians seem attracted to the seemingly hidden and forbidden realms of the dark side of the Tree of Life glyph. They exert a peculiar fascination, and appear to induce in the magician feelings of mastering powerful forces, which are represented by a complex array of symbols and *sigils*. (A *sigil* is a form of self-spell that is supposed to work on the unconscious of the magician. The magician creates a form of meaningful symbols, charges them through ritual, and then leaves them to work.) Grant gives details of the 23rd tunnel of the Qliphoth as follows:

> The 23rd kala [meaning principle of the Qliphoth] is under the dominion of Malkunofat who lies in the depth of the watery abyss . . . The key to this glyph lies in the number 61 . . . [which is] the number of Kali, Goddess of Time and Dissolution. Her colour is black, which equates her on the one hand with the void of space, and on the other with the symbolism of sexual magick typified by the blackness of gestation, the silence and darkness of the womb. Above all, 61 is the number of the 'Negative conceiving itself as Positive'. This it does through the BTN (61) or womb of Kali. BTN derives from the Egyptian *but*, the determinative of which is the vagina sign. The womb is the *nave* or NVH (61) which by metathesis becomes HVN, meaning 'wealth', the nature of which is explained by metonymy. The reference is recondite and pertains to the Goddess of the lunar serpent that appears only when it wishes to drink. It then rests its tail on the ground and thrusts its mouth into the water. It is said that 'he who finds the excrement of this serpent is rich forever'. The excrement to which allusion is made is not anal, but menstrual . . . (Grant 1994:216, 217).

Here power is associated with a Tantric view of the blackness of gestation in the womb of Kali, goddess of time and dissolution, female sexuality, death and menstruation.

The issue of the Qliphoth raises differing views among magicians. A chaos magician who was working on the Qliphoth informed me that there are pathworkings on the Qliphoth just as there are pathworkings on the other levels of manifestation of the Kabbalistic Tree of Life. He described himself as a healer, and said he was using the Qliphoth to gain control over illnesses, as each Sephirah has specific diseases that have to be mastered. This involved overcoming his specific demons and monsters, which frequently woke him in the night, and which he described as 'constructions of mind'. I had the impression that he found this aspect of magical practice both challenging and exciting, because he told me that he had left a high magic school because it was boring. However, others are much more wary. At a high magicians' workshop I was involved in a discussion with a hypnotherapist from Darlington who was in his late twenties and another man, also in his twenties, from London. During the interval an interesting conversation about the 'dark forces' arose. The hypnotherapist talked about how he was monitoring a young woman in his lodge. He had warned her not to use the Qliphotic forces, and she had promised him that she would not. He said that if he found out that she was using them he would ask her to leave the lodge. The other man told how he had been given the name Baratchial by a magician friend. He did not know what it meant, but when he looked it up in his book he found out that it was Qliphotic and he phoned his friend to find out more about it. The hypnotherapist told him to have nothing more to do with his friend.

Thus a high magician may work with the 'powers of light' or the 'powers of dark', but both practices are shaped by the Hermetic gnostic Kabbalistic system, which embraces the view that Christ and Lucifer were brothers and equally the sons of God. This position differs from orthodox Christianity, which teaches that evil was brought into the world through Lucifer's rebellion from God and is directly related to the Fall of humanity.

I now turn to contemporary witchcraft, a magical practice that also seeks to incorporate the 'dark', but that has a very different cosmology.

The Internal Source of Morality (2): Witchcraft

Modern Witchcraft: An Anti-Christian Political Movement

The idea that people who were tried for witchcraft in Early Modern Europe were peasant followers of a pre-Christian religion was first developed by Jules Michelet in *La Sorcière* (1862). He portrayed the pagan witch religion as the repository of liberty through the tyranny of the Middle Ages (Hutton 1996). Michelet re-created medieval witch rites, 'describing them with all the sympathy due from a modern liberal to an oppressed class and culture' (Hutton 1991:300–1). This idea was taken up by Charles Godfrey Leland when in 1899 he published the 'gospel of a medieval witch cult' from information supposedly given to him by a Maddelena who was a member of an allegedly hereditary group of witches in Tuscany. In Leland's myth Aradia came down to earth as a female messiah to establish witchcraft before returning to heaven. The rich had made slaves of the poor and had cruelly treated them: 'in every palace tortures, in every castle prisoners'. Aradia taught people how to destroy the evil race of oppressors when she had departed from the world. Much of modern witchcraft morality is based on Leland's myth of the Goddess Diana and her daughter Aradia. In this myth there are strong similarities to Jesus the Messiah, who is the Son of God who goes to earth to redeem a fallen humanity;

but Leland also added a strong social and political message against Christianity as the religion of the upper classes (1991[1899]:8).Thus contemporary witchcraft has been created from elements of a romantic anti-Christian political movement based on the idea of pagan 'freedom fighters'. Witchcraft thus represents liberty and freedom from oppression.

The other strand of influence that affects contemporary witchcraft notions of morality is the conception of the magical will developed by Aleister Crowley. Doreen Valiente, Gerald Gardner's High Priestess, demonstrates Crowley's legacy of 'every human being is intrinsically an independent individual with his own proper character and proper motion' (Crowley 1991:XIV) combined with a policy of 'non-harm' when she writes that: 'Witches do not believe that true morality consists of observing a list of thou-shalt-nots. Their morality can be summed up in one sentence, "Do what you will, so long as it harms none"' (1986:36).

When I asked the feminist witch Leah about the ethics and politics of the practice she replied:

> Some people are truly called to the Craft, other people do it for a variety of reasons, ego, personal gain, who knows, for sex, people have their own reasons. . . . I do believe what we do is political, and that has changed from what I meant by political ten years ago. I mean it now in more subtle ways – deep inner change of consciousness and awareness of the larger patterns.

Leah said that the emphasis was on changing the self to effect change in this-worldly reality, and that this was seen to be the basis of the attainment of power in esoteric training. Leah told me that magic was about changing consciousness; it was a way of 'changing ourselves':

> As we change ourselves we change the world. I think magic is a way of changing the world, but I'm very careful – I don't work for Labour to win the next election, for example. It doesn't start there, it starts inside. You have to be careful, because there are patterns that are dangerous to disrupt; which is why I will work for an opening up of people's hearts so people will be compassionate . . . I will not work for a piece of political legislation . . . I would work for the ranchers to see the reason why they should stop cutting down the rainforests. Change comes from within. It's very subtle.

However, practising the Craft is not an easy spiritual path, and Leah, who teaches a women's magic group, compares it with fundamentalist religions, which give clear guidance on moral behaviour:

> People are going back to fundamentalism because it tells you what to do, then you can do it. You don't have to think about it and you don't have to take responsibility for what you do. But the Craft doesn't tell you what to do – there aren't any rules, commandments,

holy books, there's nothing, you've got to take responsibility. It's not an easy spiritual path to be on. You've got to make up your own position as you go along. It's not 'let's go out and be happy together'; that's great, but it's not what it's about. It's about something much more serious. I find it very difficult to know these days whether anyone's into it [fundamentalism], because they are into surviving and paying the mortgage, getting their kids into a good school and not getting AIDS.

I find it more and more difficult to say anything definite. How dare I say anything definite about this realm? The way I approach my path is quite different to most witches. I try to look at everything on a larger level; then I am less able to make a clear statement about anything – which is why teaching a women's group is difficult, because they all want answers and I don't have any answers . . .

Organic Totality: Influencing the Flow

Morality in high magic is largely a clear-cut affair – although there are different perspectives on working the Tree of Life glyph, either taking the Christian 'Light' approach, or Aleister Crowley's anti-Christian rebellion towards the 'dark', both systems offer a clear cosmology and structure based on the Judaeo-Christian system of dualistic opposition between good and evil. In the Christian tradition of the Hermetic Order of the Golden Dawn the emphasis was on the individual, the internal battle between good and evil within the self and the battle for redemption from 'the Fall'. It is thought that by improving the magical self the cosmos can spiritually evolve too. When Dion Fortune and her magical Fraternity were channelling the forces of good against the forces of evil during the Second World War she was not only drawing on notions of difference between group souls of race, but also using the Christian notion of the dualism of the forces of good and the forces of evil that arose after the Fall from an original primal and harmonious state. In contemporary witchcraft, however, the situation is made more complex by virtue of its monistic framework. Although contemporary witchcraft developed from high magic and shares many of its ideologies, the biggest difference is cosmological, for the witchcraft view of the cosmos is as an organic totality.

It is possible to draw comparisons, and some witches do, with the Taoism of the Chinese philosopher Lao Tzu. Tao is a dynamic, vital force with innate powers of potential (Cooper 1981:13); it is also the ultimate reality and energy of the universe – the Ground of being and non-being (Watts 1981:40). It is similar to the early Greek worldview of Thales, Anaximander and Anaximenes that appeared to view nature as an intelligent organism, on the basis of an analogy between the world of nature and the individual human being as microcosm (Morris 1994:25–6). There is no division between spirit and matter or god and nature. Witchcraft cosmology is said to be life-affirming and in harmony with the ordering principle of the cosmos. However, witchcraft goes farther than Taoism, which is based on a positive non-intervention with the Tao – the life force. Witchcraft practitioners attempt to influence 'the flow' of the cosmic forces. According to Ken Rees, the sociologist and wiccan, wicca falls between Taoism and practices influenced by Aleister Crowley that strongly emphasize the magical will. He believes that wicca is a combination of going with the flow and influencing it at the same time. Rees told me that wiccans intervene in the flow not typically for themselves alone but also as a service to others, and that the role of the wiccan priest or priestess was to serve the community, especially through the creation of sacred space. For Rees, this vocation is not twenty-four hours a day, but just within the ritual circle. He makes a clear distinction between ritual space as a powerful and specially prepared centre for visualization – he called it an 'amplification chamber' – and the ordinary mundane world. He told me that an early wiccan conception of the ritual circle was of a dynamo, a space where power was heightened and 'had more chance of working'. According to this view, a magician becomes a channel for the universal forces and energies (represented by gods and goddesses) and navigates the time–space continuum within the ritual circle, which is often termed *wyrd*. *Wyrd* in Anglo-Saxon means 'destiny', but also 'power' or 'prophetic knowledge'; it is a term used to express the magical interconnectedness of all life. It has been popularized by the work of the psychologist Brian Bates (1996a[1983], 1996b). In the Northern tradition journeys in sacred space are an opportunity to experience 'time-flux', and are seen as adventures rather than aimed at reaching a destination; they express the power of an individual to negotiate their own life pattern within the larger web of the universal Wyrd.

According to this view, wiccans are priest and priestesses of sacred space, and are thereby bound to the 'laws' of the cosmic forces. This is demonstrated by ritual symbolism: the ritual paraphernalia of wicca represents the magicians' relationship with wider cosmic forces. One coven with which I was practising had a black altar cloth embroidered with silver moons (one at a different phase at each corner) and a red pentagram in the centre that they used for *esbats* (moon rituals). Two cut glass bowls with pentagrams in the bases were placed on top and were filled with salt and water for purification, and a wooden wand encrusted with jewels at one end was placed in the centre of the cloth next to figures of Goddesses and Gods. The red pentagram in the centre represented the microcosm of the magician: the four elements of air (mind), fire (will), water (emotions) and earth (body) plus spirit. The microcosm of the magician is in correspondence with the phases of the moon. The salt water is used to purify the sacred space and also the magicians present, and finally, the wand represents the magician's will aligned with cosmic forces. When a circle has been cast, the space inside is seen to be sacred and 'between the worlds'; it is a place where participants in the ritual commune with the gods, and everything that happens within that space is seen to be in accord with cosmic and divine forces.

At the end of one ritual with a wiccan coven, before the circle was closed, a discussion developed about sacred space. Food and wine were shared, and the coven members were very relaxed. One wiccan priest suggested that we should show how sexually liberated we were by performing the actual Great Rite. This started an intense conversation about wicca and sacred sexuality. The high priest told us about how, as an initiate, he saw his role as a teacher, and he said he wanted to teach skills in sacred sexuality. I asked if people had to undergo a change in consciousness in order to perform sexual rites, and if so, whether this was created by the sacred space of the circle. The high priest told us that the wiccan path offered everything that anyone could want from magic – sex, drugs and all the ultimate experiences – and that if it was conducted in sacred space then it was holy. A discussion ensued about the boundary between sacred space and everyday reality. One woman spoke about how what was seen to be liberatory for men was not necessarily liberatory for women and drew a parallel with the development of the contraceptive pill and the 'free love' of the 1960s, which put pressures on women to behave in a certain way. Another point raised was that of power dynamics within groups, and how the introduction of sex brought in a highly dangerous area where power could be manipulated. The coven did not reach a consensus, but the notion of sacred space – as an 'automatic' ethical area outside the constraints of ordinary space, time and morality – was an important area for discussion. At some point later in my fieldwork, another wiccan told me of a coven he knew that practised 'black magic' through the high priest's abusing his position. This wiccan told me that he thought that people abused power through sexuality, and that this was negative, and therefore black magic; but he thought that we 'need the negative and the black in Paganism because it creates balance'. He added that, 'to know what is good and bad is to know the difference – without it everything would be the same'.

Witches frequently work with the 'dark'. The witches I have spoken to clearly feel that this particular aspect of their magical work is very important, and point out how vital the dark is to their practice and identity: the dark must be honoured. Many told me how they were working with forces on the dark phases of the moon, and how this raised profound and deeply emotional issues. Witches work magical rituals at night with sexual energy and the deep emotions that this raises – this practice is seen as the source of much magical power. Such a power Christianity has labelled as 'evil', and is constantly working against in its dualistic cosmic battle of light against dark.

Witchcraft rituals may be performed for healing (see Chapter 5), or, as one wiccan explained to me, they may be enacted for 'the grey area of magic' – 'hexing', 'sending' and 'fetching' energy for a specific purpose, or 'binding'. This form of magical working is often based on spell-making, whereby the coven will raise energy that is 'sent' into the 'ether' to work by intent to achieve a given end or set of circumstances. Spells, defined as rituals for raising psychic power that is directed to a specific and practical purpose (Farrar and Farrar 1991:235), are an ambiguous area, because they are an attempt to determine the future – to influence the cosmic flow; but they are also said to 'speed up destiny' and to work on the potential that is already there. They are seen to be capable of 'going either way': i.e. they may not work in the manner intended, and may have unintended consequences.

Hexing has to be done carefully, as it is said that hexing the innocent returns ten-fold. Leah said:

> I am of the tradition which says that you cannot heal unless you can hex. I can do both. I can choose one over the other, but my reasons had better be good, because everything you do goes out into the cosmos. If you do something that is skewed then that is the energy that goes out to the cosmos for ever, and it goes on and on like Voyager. And I've had that happen. And so you create the larger reality. A huge responsibility. You must have ethics.

Thus the practice of witchcraft places great responsibility on the practitioner. The Dianic witch Z. Budapest in *The Holy Book of Women's Mysteries* says that if women hex rapists or others who 'commit crimes of patriarchy', there is no divine retribution. She gives instructions on how to perform a 'Righteous Hex', for 'violent criminals only' and for when you 'know, not just think' that 'someone has harmed you' (1990).[4] But I have heard mention of hexing being done between witches for more mundane reasons, over quarrels about money for example, or to gain retribution against an employer who was unsympathetic.

Witches may also 'bind' or curse. I was a member of one coven that spent a long time deciding whether to 'bind' a fellow witch because she was seen to be manipulating power. Binding is seen to render the offender powerless rather than to punish them, which will happen anyway through Karma. In the event the group disbanded rather than taking any positive magical action.

Another wiccan spoke to me about a coven of which he was a member. The high priestess had spontaneously turned a standard dark moon ritual into a magical space to send a curse to a ritual magician who had been physically molesting her. I asked about the ethics associated with this, and whether he was implicated in the process. He said that although he had participated in the energy-raising the responsibility lay with the sender of the curse. He said that if he had objected to the magical working he would have been able to sit out and no one would have thought any less of him. This wiccan said that he also 'set up the circumstances' whereby someone could curse in a ritual circle on the basis that the 'unjustified comes back to the sender'. He thought that this was ethically justifiable as long as the reasons for the curse were fully discussed beforehand.

At the feminist witch workshop on class and sexuality that I described in a previous chapter the emphasis was on going down into the dark. The workshop was held in October, and Samhain (Halloween) was approaching. On the initial meeting on the Friday evening, we sat in a circle around the cauldron and spoke in turn about how we felt about the dark time of the year. It was a time for the women to look inside themselves and face issues concerning the deep meanings in their lives – of growth, death and transformation. Some spoke with trepidation; some welcomed the change from the summer, which was seen as a time to be more outgoing and expansive. We went down to the Underworld on that first evening using the myth of Inanna, and most said that they had found it a moving and powerful experience.

Thus the high magical conceptual frameworks of good and evil are largely redundant in witchcraft. However, as I have shown, witches do work with what they term light and dark forces. Like those high magicians who focus on the dark, they believe that there has been too much emphasis on the light, and they seek to redress the balance. Consequently much ritual work focuses on the dark moons and deities associated with the dark or the underworld. I have spoken with many witches about issues of morality and 'evil', and all have shared the common sense of inappropriateness of the term 'evil' in relation to their practice. They say that they prefer to integrate what they see as 'positive' and 'negative' forces. In contrast to high magicians, they do not use the notion of the Devil or Satan either, seeing these as inventions of Christianity. However, on a practical level the notion of evil is still used in a manner common to many other religious groups: for example, an informant told me how she had joined a particular coven and had problems with the High Priest, who she felt had inflexible views and was telling her what to think. When she decided to leave he was angry, and she could feel his 'bad vibes' for some time after. She felt that he was still trying to control her. She eventually confronted him, and the vibes stopped for her. But he maintained his intransigent stance and instructed 'his' coven not to speak to her or visit her bookshop, as she was 'bad'. As so often happens, evil in the shape of dissenters is frequently expelled, rather than being brought into balance, as the ideal theory would have it.

Ideologically, Paganism is said to be a tolerant spiritual path; but a Pagan Federation regional meeting that I attended turned out to be a cauldron of recriminations and discontent. This meeting was convened at the annual Pagan Federation Conference to formulate questions to pose to the Committee. One person said that he could never get anything published in the PF journal, and he felt that his views were not approved of (he described himself as a Pagan chaos magician). Another woman, who organized many talks and events, said that she could never get her meetings advertised. She felt that this was because she was seen as competition, and therefore a threat to their organization and views. Someone else who was associated with the Progressive Movement, a group of magicians who had broken free from what they perceived to be the dogma of 'traditional' hierarchical witchcraft, felt that he had been excluded by the PF Committee because 'his face didn't fit'. It was suggested that some members may be 'blacklisted' on the computer, and some were judged as not respectable. As the group was breaking up, one woman spoke of how disappointed she was to hear all that had gone on: 'I didn't expect this of Pagans', she said with feeling.

The Wiccan Rede: Individual and Community

As I noted in Chapter 4, one of the biggest differences between high magic and witchcraft is centred on the issue of community and locality of place – witchcraft is seen to be a nature religion. Witches, as Pagans, are supposed to have an association with place – the tangible everyday world. This notion is less well defined in high magic, which tends to emphasize spiritual or cosmic evolution rather than nature as such. Connection with community and environment, together with individuality and freedom of will, is central to witchcraft; but this particular combination or Pagan ethos is problematic, and has prompted an ongoing discussion in Pagan newsletters and magazines. One contributor to the *American Circle Network News,* writing on Pagan ethics, suggests reflection and meditation upon an individual's own system of ethics: 'Your choices may be some of the same things I have chosen, or they may not. One of the best things about Paganism is its individuality.' This writer noted that a balance was needed between individualism and communal ethics. He pointed out areas that Pagans could think about. Firstly, the value of community service and the standing up for beliefs such as 'Think Globally, Act locally' by political activism; secondly, the changing of harm in the individual's own life; and finally, to stop the harm the invividual is causing. 'Figure out how you are causing harm, and do your best to stop it . . . the Lord and Lady . . . put us on this good Earth to spread joy and love . . . Causing harm is antithetical to spreading joy and love. Love yourself first, and everything else will fall into place' (*Ashley Ravenwood Circle Network News Spring*, 1996, Vol. 18: no.1, p. 19).

In modern witchcraft, the Wiccan Rede is seen as a positive restriction. Thus witchcraft ethics are seen to be positive rather than restrictive, based on 'blessed is he who' rather than 'thou shalt not' (Farrar and Farrar 1991:135). 'Do what you will, an it harm none' is based on the idea that power comes from within and that truth must be found inside. However, the Wiccan Rede is a problematic basis for the creation of morality in everyday life; it is difficult to interpret practically. Kari Scott, another contributor to the *American Circle Network News*, finds it vague and passive. She makes three important points. Firstly, while it reflects Paganism's 'staunch individualism', it is inadequate because Pagans fail to question the Rede 'for fear of creating dogma'. Secondly, discussions on how to follow 'An it harm none' remain scarce. Thirdly, the passive nature and broad interpretation fail to provide an adequate ethical structure for Paganism as a community. Kari Scott argues that the repudiation of anything even remotely resembling 'dogma' for fear of inhibiting freedom means that the Rede itself becomes dogmatic, because it is accepted as truth without question. In addition, it is unclear how a Pagan moves from the Rede to action in everyday life. She argues that most Pagans accept eco-centrism and egalitarianism as examples of the Rede at work. They claim to live up to these interpretations of 'an it harm none' by preserving the Earth through recycling, voting for environmental protection legislation, and by respecting both women and men from a variety of religious traditions and cultures. However, the lack of discussion on ethics suggests otherwise. In addition, she suggests that the Rede may be interpreted broadly, suggesting that she would view racism and sexism as harmful, but others might not. They might limit the Rede to physical or magical harm, and see verbal disrespect as ethically acceptable. The writer suggests that Pagans must find a balance between staunch individualism and communal ethics: 'Through philosophically exploring "Pagan ethics", we can create a moral system which protects all from a well-defined "harm" and encourages positive community action' (*Circle Network News*, Spring 1996, Vol. 18; no. 1, p.19).

The tensions between the commitment to inner truth and the reality of the collective endeavour were brought out clearly in an interview with Jane, the feminist witch. She told me about the problems she had experienced with community. These problems became increasingly more serious for her over a period of several years. On reflection, she felt that she not been able to fit in with the unspoken 'code of conduct' in the group. The coven ideology was based on the Reclaiming Collective in San Francisco, but she said there was also an implicit guru-type relationship with the Collective, which made it difficult for anyone with a clear sense of their own centre (of who they were – their own truth), as this created contradictions and misunderstandings. Jane said that real community 'never got off the ground', because they were all there with their own agendas, which were submerged under the unspoken group consensus, which was strongly influenced by certain key members of the Reclaiming Collective. Part of this consensus was shaped by the 'Twelve Step Programme', where talking and being intellectual were strongly discouraged, because 'You are supposed to find and express your feelings.' This had the effect of putting a 'very low value on thinking for yourself and asking questions'. Being a coven member required learning how to fit in with and sense the required behaviour from those in the British coven who had a connection – a hot-line – with the gurus in San Francisco. The true route into the coven was fitting into the unwritten agreements – of adapting and learning to come up with the right words (this is similar to my experience of high magic training – see Chapter 3). Nothing was clearly stated, but what was required and appropriate behaviour 'filtered through like osmosis'. People were not encouraged to be themselves, but to fit into a fantasy of community, which Jane thought was very unhealthy. Members of the coven had a catch-phrase of being 'honest and open', which they were always talking about, but never really practised. If Jane attempted to bring up some issue she was either not listened to or encountered rage, 'cut off', tantrums or emotional manipulation. On various occasions she experienced her behaviour's being pathologized, and once she was given a self-therapy book and asked to look up her symptoms so that she could 'help herself' and therefore 'sort herself out'. Jane reluctantly concluded that she had to accept the possibility that her coven members were very disturbed people, and that magic was a 'natural alternative' to drugs and alchohol; in theory it was a spiritual path, but in practice it was based on escapism and fantasy, and did not deal with the issues and problems of the everyday world.

Jane had recourse to the otherworld for guidance on how to deal with the difficulties with her coven. She explained that on the previous Saturday she had done a journey with her shamanic drumming group, and one of her power animals had told her to go and talk to the fairies. So on the Sunday she went to a local wood, where she did personal ritual work:

> I visited various parts of the wood to 'feel' the right place, and eventually settled on one. I walked slowly in a circle, silently calling on the spirits of the place, and 'feeling' for the appropriate action to take. I sat on a fallen silver birch trunk, in front of a beech tree, and silently called on the faeries to tell me what I needed to know. I sat quietly for a while, when I saw a vision of a mirror in front of me, slightly at an angle, so I could see faces of people as they were standing behind me . . . I knew them all. Then the mirror moved so that I could see my own face, and then back to them again.
>
> The message was that I should reflect on what were my issues, and what were their issues, that I should reflect their issues back to them, and deal with my own issues myself. When I came back to myself, I walked in a circle in the opposite direction, thanking the space, the spirits and the faeries for helping me. I then went for a long slow walk in the wood before going home.

Jane's experience clearly demonstrates the problems of working in intimate small ritual groups and the difficulties attendant on 'finding your own truth' or sense of morality within and being able to live it. It also shows how the otherworld can be used as a source of knowledge on how to deal with problems such as Jane faced. Jane has not been able to talk about these issues with her group.

A Comparison: High Magic and Witchcraft

It can be seen from the foregoing that high magic and witchcraft (wicca and feminist) are intimately connected: they share the same roots, and both have an internal sense of morality guided by intuition and communication with otherworldly forces. They also share a common notion of Karma, as a moral system oriented around good and non-good action and its consequences for future rebirths – an action and its result have an effect in the subsequent life of an individual – and this structures notions of morality. The role of Karma is to teach spiritual lessons to attain harmony (Fortune 1987a:17). In high magic virtuous actions and spiritual practice aid spiritual evolution. These ideas are generally shared by witches, but there they are much more vague and do not form any specific cosmology of their own; they are not worked out in a systematic manner, but are encompassed in a variety of implicit and explicit values.

Yet high magic and witchcraft have different cosmologies and ideologies that determine their conceptions of morality. This difference affects the way that magicians view the world and the way they practise magic. A comparison of the two will illuminate important differences:

High Magic	*Witchcraft (wicca and feminist)*
Strong moral or anti-moral framework based on esoteric Christianity or anti-Christianity	The Wiccan Rede
Initial dualistic worldview based on 'the Fall' and a battle between 'good' and 'evil'	Monistic worldview; the world is intrinsically good and harmonious
Spiritual evolution	Practical: healing and hexing
Individual as microcosm/star	Individual in relation to community and nature

As Tipton (1982) has pointed out, alternative religions tend to lay out a relatively detailed picture of reality that is analytically complete in its own terms and based upon ritual experience. This supports an explicit, unified ethic that gives coherent answers and clear ideas about true selfhood that concern union with the infinite. In high magic this is symbolized by the Light. The emphasis in the Christian version of high magic is on moral perfection, which has the effect of setting practitioners against the wider society as 'guardians' of morality and goodness in the battle of light against dark forces as demonstrated in the work of Dion Fortune and her magical 'battle of Britain'. God's will is not fixed, explicit or knowable through sacred scripture, as in mainstream Protestant Christianity; rather it becomes known through the internal dialectical battle between good and evil – evil being used for the greater good. Aleister Crowley's version of high magic also provides a clear framework (both use the Kabbalistic Tree of Life glyph) because it is based on the reversal of Christian light and an exploration of the negative, the Qliphotic, the moral otherside of Christianity – the evil or dark forces – in a similar quest for gnosis while also incorporating notions of autonomy and individualism.

Magic and Morality: 'A Mixed Spectrum'

By contrast, witches search for moral coherence in the resolution of conflict between positive and negative forces within the harmony of an organismic conception of cosmos. As I have argued, this is inherently problematic, because Aleister Crowley's maxim of autonomy and the magical will does not offer a clear-cut moral framework for practising magic in a community and environmental context. In addition, witchcraft covens are autonomous small units that interpret the 'flow of the cosmos' with its various forces in different ways, and there are no absolute 'rules' guiding spell-making other than personal responsibility and the veiled threat of an unjustified spell's coming back threefold. The difference between witchcraft and high magic concerning wrongdoing may be located in ideas about the sources of wrongdoing. High magic sees evil as internal, whereas witches are more inclined to deal with 'negative forces' through hexing in the here and now. This contrast can be illustrated by Martin Southwold's useful analogy, which draws on Lienhardt's (1961) study of the Dinka to compare external and internal explanations of occult workings. The Dinka tend to project on to supposed external occult agencies much that we attribute to the inward workings of the mind, whereas in Buddhism, for example, the tendency is to reduce all occult agencies to inner psychic phenomena (1985:139). In both Buddhism and Hinduism 'evil' is accommodated within a unified existence (Löwith 1967; Southwold 1985). Southwold notes that Buddhists are taught to see the roots of wrongdoing within the self; whereas in other cultures these are blamed on external occult agencies (1985:139). In a similar fashion, Lionel Caplan, in a study of Protestant Christians and Hindus in Madras, makes a distinction between evil that is humanly motivated, such as sorcery, and the personification of evil as occult forces beyond human control. In the latter case evil is projected outwards, and wrong-doing is attributed to non-human occult agents (Caplan 1985).

High magicians take on the moral responsibility for all the evil in the world, as my magical supervisor explained to me, and this increases their sense of duty – a duty that is on the macro-cosmic rather than a local and community level, as in witchcraft. In addition, the role of Karma in high magic is viewed in macro-evolutionary rather than practical everyday terms: it is concerned with spiritual evolution in the cosmic sense and internal evolution seen in terms of the transformation of the lower self to the higher, as in the Abra-melin ritual.

It is clear from the foregoing that there are only so many ways that a religion can organize notions of morality, and high magic and witchcraft offer two variations within this broader scheme. Morality is ultimately individual and is defined by 'cosmic law' rather than by reference to the wider society. In practice notions of morality often form a grey area, what one magician described to me as a 'mixed spectrum', rather than 'white', 'black', 'good' or 'bad', because ultimately all reference is made internally and, as the feminist witch Jane's experience has shown, unspoken and implicit group assumptions may effectively stifle 'internal truth' as self-knowledge. This corresponds closely with my own experience of participant observation training in a high magic school. To progress within the school or fit into a magical group the individual magician often has to learn the required response. It seems inevitable that much power in magical groups stems rather from this-worldly manipulations by some individuals in pursuit of a status and an identity that they present as legitimated and morally sanctioned by higher powers.

Notes

1. It was interesting to note that, in a recent Divine Magic television series shown in January 1996, this Magus, who was described as a 'Voodoo Priestess', took a journey back to Haiti and was initiated into a Voodoo group.
2. Given that this was a time when allegorical myths were popular, the question of Abraham the Jew's physical reality is questionable.
3. Mathers notes variations in the text ranging from Abramelin, Abramelim and Abraha-Melin (MacGregor-Mathers 1976:xxii).
4. Z. Budapest recomends that a black altar be constructed and a doll-shaped form resembling the enemy be cut out:

> . . . sew it around from east to north to west to south (widdershins), and leave only a small part open. Stuff it with boldo leaves . . . Indicate the eyes, mouth, nose, and hair on the doll. On a piece of parchment paper, write the name of your enemy and attach it to the image.
>
> [The Goddess Hecate is invoked to doom the life of the enemy:]
>
> *Goddess Hecate, to you I pray,*
> *With this enemy no good will ever stay.*
> *Cut down the line of his life in three,*
> *Doom him, doom him, so mote it be!*
>
> When you pronounce this, take a mallet and break his 'legs' by breaking the herb inside. Dust it with Graveyard Dust; anoint it with Double Crossing oil, and burn your Black Arts incense. Imagine him totally miserable and with one leg broken . . . Do this three nights in a row. On the third night, burn the doll and bury it. Draw a triple cross over his grave with Dragon Blood power. Walk away without looking back . . .

Note: Dispose of hexes as far away from your house as possible. Each night you can break something else in him, or stick black-headed pins into his liver or penis. May patriarchy fall! (1990:43).

—8—

Conclusion

For magicians the otherworld reaches into this world through organized rituals, pathworkings, guided meditations, myths and stories; in theory it is part of this world, the boundaries between the two domains are permeable. Magical energy flows between like osmosis, linking everything in a cosmic dance, an encompassing totality of forces both benign and malevolent. The otherworld is co-existent with everyday reality, and is not supposed to negate this world, nor is it a cathartic fantasy land (although it may be). The otherworld is a magical realm that is hidden from the everyday world, and only some are privy to its delights, as Doreen Valiente (1978) expresses it in her poem 'The Witch's Ballad':

Oh, I have been beyond the town,
Where nightshade black and mandrake grow,
And I have heard and I have seen
What righteous folk would fear to know!

For I have heard, at still midnight,
Upon the hilltop far, forlorn,
With note that echoed through the dark
The winding of the heathen horn.

And I have seen the fire aglow,
And glinting from the magic sword,
And with the inner eye beheld
The Hornéd One, the Sabbat's lord.

We drank the wine, and broke the bread,
And ate it in the Old One's name.
We linked our hands to make the ring,
And laughed and leaped the Sabbat game.

Oh, little do the townsfolk reck,
When dull they lie within their bed!

Beyond the streets, beneath the stars,
A merry round the witches tread!

And round and round the circle spun,
Until the gates swung wide ajar,
That bar the boundaries of earth
From faery realms that shine afar.

Oh, I have been and I have seen
In magic worlds of Otherwhere.
For all this world may praise or blame,
For ban or blessing nought I care.

For I have been beyond the town,
Where meadowsweet and roses grow,
And there such music did I hear
As worldly-righteous never know.

Valiente makes a clear distinction between the town – the everyday social world with what she sees as its dull moral righteousness – and the far hilltop, associated with the witches' plants deadly nightshade and mandrake, the haunt of a group of laughing heathens. A circle is created, a reversal of the Christian Mass is celebrated, and the 'inner eye' of magical awareness enables the gates of the otherworld to swing wide open, revealing the shining faery realms *for those able to see.*

   Because of the way that Western cultures have developed, this dualism between the dull ordinary world and the enchanting other realms has been enforced and maintained by both magical practitioners and mainstream culture. For magicians, otherworldly reality often has more reality and importance than this-worldly reality, as was demonstrated during my high magic training, when I was told to let go of all my belief systems. All socio-political problems were interpreted (and solutions explained) by recourse to the otherworld, and if humanity destroyed the planet it would be given another chance. In addition, otherworldly status does not necessarily translate into everyday reality, and this maintains a strict demarcation between the two realms. Women are venerated in most magical practices (especially witchcraft); but it does not necessarily follow that high evaluation in the otherworld translates into equal status for women in the ordinary world. The religious aspect of worship does not equate with changing the social world; indeed it frequently reinforces it, giving gender stereotypes romantic or even divine legitimation.

   Mainstream culture often views magic with suspicion, treating it as 'the occult', the irrational, or sanitizing and sweetening it, as in the films of Disney. Within the discipline of anthropology the otherworld has been neglected as an ontological source, and attention has been focused instead on the social parameters of magical practice. By contrast, this work has attempted to show how the otherworld shapes contemporary Western magicians' lives and is vital to an analysis of magic. Communication with what is seen as a powerful otherworld is central to the practice of Western magic. Magicians' experiences and expectations of magic are very different – there are a variety of approaches and traditions; but the thread connecting the disparate groups is their connection with an otherworldly reality. By using my own subjective experience of magic I have focused on this otherworldly realm of spirit, which, as a result of our Western cultural history, has been denigrated and marginalized as a source of knowledge. In the process I have sought to break down the boundary not only between this world and the otherworld, but also between anthropologist and magician – researcher and researched. I believe that it is only by breaking down these barriers that a full interpretation of magic is possible; it is only through an engagement with the magical imagination that an analysis of the relationship between different worlds is possible. Being an anthropologist in such a research situation has been an uncomfortable process; it has involved self-examination and exploration, often in a confrontation with fear; but it has been conducted within a broader commitment to further understanding and communication.

The issue of power has been a central preoccupation in this study, both within the magical subculture and in the anthropological encounter. I have shown how the acquisition of a magical identity concerns spiritual healing – a 'reconnection' with the otherworld as the true source of being and empowerment. The control of otherworldly powers can bring personal power; but its misuse is also said to bring imbalance and, at worst, madness. The otherworld is the ultimate locus of legitimation and authority: how does this affect social relationships in everyday reality? It is said that the final arbiter of otherworldly communication is individual experience; but this is belied by the charisma of certain individuals and also the traditions of particular occult schools and various witchcraft practices. While acknowledging the value of training and discipline, there is an unacknowledged fundamentalism – an over-reliance on the words of magical adepts or experts – and a pressure to conform in many magical groups, despite an ideology to the contrary. The origins of particular venerated interpretations of the otherworld stem from individual vision – be it that of Aleister Crowley, Dion Fortune, Gerald Gardner or Starhawk. Like more organized religion, a certain dogma – which masquerades under the guise of individual freedom to experience – surreptitiously controls both further interpretation and group dynamics. Issues of power are not openly discussed, and the frameworks for practising and understanding the otherworld are rarely questioned. Thus the otherworld is a central source of conceptual and experiential knowledge through which power relationships are negotiated. Otherworldly connections with spirits or gods (however they are personalized) are used to legitimize individual power relationships. Some magicians' spiritual 'connections' are stronger, mightier or more 'authentic' than others': thus the otherworld is used to maintain this-worldly forms of control over others.

Despite this, for many magicians being intimately involved with the otherworld is an empowering and positive experience, one that is lovingly explored through mythologies of gods and goddesses, stories and dreams. It is approached through the imagination and a language of metaphor and analogical thinking. This work is a call for a common language – or at least a neutral space (if such a thing exists) – to analyse magic as a part of human experience. A perspective of reality in terms of the otherworld brings another way of being into view – one that can enrich our understanding of what it is to be human and all the complexities that that entails. The realm of spirits has a cosmological unity, and its magic is rational, as Evans-Pritchard suggested more than fifty years ago; but this rationality should not be opposed to science. Samuel's theoretical framework offers a different vision of science that incorporates both formal and informal (or non-scientific) knowledge, and is broad enough to include experience of magical otherworlds. This study is a part of a growing awareness that our scientific conceptual frameworks need expanding. The bringing together of two previously opposing theories – in this case magic and science – suggests that a new way of explaining the world is possible. Instead of opposing magical thinking and scientific thinking, a dialectical interplay between the two – through a study of multiple ways of knowing and their attendant power relationships – could help us to approach the subject of magic in a constructive manner. The crucial issue is that such questions should be asked by the scientific discipline of anthropology.

Bibliography

Adler, M. 1986 [1979]: *Drawing Down the Moon.* Boston: Beacon Press.

Airaksinen, T. 1995: *The Philosophy of the Marquis de Sade.* London: Routledge.

Ankarloo, B. and G. Henningsen, eds. 1993: *Early Modern European Witchcraft.* Oxford: Clarendon Press.

Anthony, D. and T. Robbins 1978: 'The Effect of Detente on the Growth of New Religions: Reverend Moon and the Unification Church', in *Understanding New Religions*, ed. Jacob Needleman and George Bakeir. New York: Seabury.

Anthony, D. and B. Ecker, 1987: 'A Framework for Assessing Spiritual and Conscious Groups', in *Spiritual Choices: The Problem of Recognizing Authentic Paths to Inner Transformation*, ed. D. Anthony, B. Ecker and K. Wilber. New York: Paragon.

Ardener, S. 1987: 'A Note on Gender Iconography: The Vagina', in *The Cultural Construction of Sexuality*, ed. P. Caplan, pp. 113–42 London: Tavistock.

Ashcroft-Nowicki, D. 1983: *The Shining Paths.* Welllingborough: Aquarian.

—— 1991: *The Tree of Ecstasy.* London: Aquarian.

Assagioli, R. 1973: *The Act of Will: Self-Actualization Through Psychosynthesis.* London: Aquarian.

Avineri, S. and A. de-Shalit 1992: *Communitarianism and Individualism.* Oxford: Oxford University Press.

Baring, A. and J. Cashford 1993: *The Myth of the Goddess.* London: Arkana.

Barker, E. 1987a: 'Brahmins Don't Eat Mushrooms: Participant Observation and the New Religions'. *L.S.E Quarterly*, June: 127–52.

—— 1987b: *New Religious Movements and Political Orders.* Canterbury: University of Kent.

—— 1992: 'Authority and Dependence in New Religious Movements', in *Religion: Contemporary Issues,* ed. B. Wilson, pp. 236–89. London: Bellow Publishing.

—— 1993: 'Charismatization: The Social Production of "an Ethos Propitious to the Mobilisation of Sentiments"', in *Secularization, Rationalism and Sectarianism*, ed. E. Barker, J. Beckford, and K. Dobbelaere, pp. 180–201. Oxford: Clarendon Press.

Baroja, J. Caro 1993: 'Witchcraft and Catholic Theology', in *Early Modern European Witchcraft*, ed. B. Ankarloo and G. Hennningsen. Oxford: Clarendon Press.

Barstow, A. 1994: *Witchcraze: A New History of the European Witch Hunts.* San Francisco: Pandora.

Barthes, R. 1993 [1982]: 'Myth Today', in *A Roland Barthes Reader,* ed. Susan Sontag. London: Vintage.

Barton, B. 1990: *The Church of Satan.* New York: Hell's Kitchen Productions.

Bates, B. 1996a [1983]: *The Way of Wyrd.* London: Arrow.

—— 1996b: *The Wisdom of the Wyrd.* London: Rider.

Beckford, J. 1984: 'Holistic Imagery and Ethics in New Religious and Healing Movements', in *Social Compass*, 31.2–3:259–72.

—— 1985: *Cult Controversies: The Societal Response to the New Religious Movements.* London: Tavistock Publications.

Bell, C. 1992: *Ritual Theory, Ritual Practice.* Oxford: Oxford University Press.

Bem, S. 1974: 'The Measurement of Psychological Androgyny', *Journal of Consulting and Clinical Psychology* 42: 155–62.

Bender, B. 1998: *Stonehenge: Making Space.* Oxford: Berg.

Ben-Yehuda, N. 1992: 'Witchcraft and the Occult as Boundary Maintenance Devices', in *Religion, Science and Magic*, ed. J. Neusner, E. Frerichs, and P. McCrackern Flesher. New York: Open University Press.

Blavatsky, H. 1888: *The Secret Doctrine.* London: Theosophical Publishing House.

—— 1988 [1877]: *Isis Unveiled*, Vol. 1. California: Theosophy University Press.

Bloch, M. 1986: *From Blessing to Violence: History and Ideology in the Circumcision Ritual of the Merina of Madagascar.* Cambridge: Cambridge University Press.

—— 1992: *Prey into Hunter: The Politics of Religious Experience.* Cambridge: Cambridge University Press.

Bloom, W. 1991: *The New Age.* London: Rider.

Boddy, J. 1993: 'Managing Tradition: "Superstition" and the Making of National Identity among Sudanese Women' (paper delivered to ASA Conference Oxford).

Bourguignon, E. 1979: *Psychological Anthropology: An Introduction to Human Nature and Cultural Differences.* New York: Holt, Rinehart & Winston.

Bowman, M. 1995: 'The Noble Savage and the Global Village: Cultural Evolution in the New Age and Neo-Pagan Thought', *Journal of Contemporary Religion.* Vol. 10 no. 2.

Briggs. J. 1970: *Never in Anger: Portrait of an Eskimo Family.* Cambridge, MA: Harvard University Press.

Buckley, T. and A. Gottlieb 1988: *Blood Magic: The Anthropology of Menstruation.* Berkeley, CA: University of California Press.

Budapest, Z. 1990 [1980]: *The Holy Book of Woman's Mysteries.* London: Hale.

Burke, P. 1992: 'We, The People: Popular Culture and Popular Identity in Modern Europe', in *Modernity and Identity*, ed. J. Friedman and S. Lash, pp. 292–305. Oxford: Blackwell.

—— 1993: 'The Comparative Approach to European Witchcraft', in *Early Modern European Witchcraft,* ed. B. Ankarloo and G. Henningsen. Oxford: Clarendon Press.

Bibliography

Burridge, K. 1967: 'Lévi-Strauss and Myth', in *The Structural Study of Myth and Totemism*, ed. Edmund Leach. [Association of Social Anthropologists Monographs, No. 5] London: Tavistock.

Butler, J. 1990: *Gender Trouble: Feminism and the Subversion of Identity.* New York: Routledge.

Caplan, L. 1985: 'The Popular Cult of Evil in Urban South India', in *The Anthropology of Evil*, ed. David Parkin. Oxford: Blackwell.

Caplan, P. ed. 1987: *The Cultural Construction of Sexuality.* London: Tavistock.

Caplan, P. 1993: 'Introduction' to *Gendered Fields: Women, Men and Ethnography*, ed. Diane Bell, Pat Caplan and Wazir Karim. London: Routledge.

Carr-Gomm, P. 1991: *The Druid Tradition.* Shaftesbury, Dorset: Element.

Carrithers, M., S. Collins, and S. Lukes, eds. 1985: *The Category of the Person: Anthropology, Philosophy, History.* Cambridge: Cambridge University Press.

Carroll, J. 1987: *Liber Null and Psychonaut.* York Beach, ME: Samuel Weiser.

—— 1992, *Liber Kaos and the Psychonomicon.* Privately printed.

Castaneda, C. 1970: *The Teachings of Don Juan: A Yaqui Way of Knowledge.* Harmondsworth: Penguin.

—— 1973: *A Separate Reality: Further Conversations with Don Juan.* Harmondsworth: Penguin.

—— 1975: *Tales of Power.* London: Hodder and Stoughton.

Cavendish, R. 1975: *The Powers of Evil.* London: Routledge and Kegan Paul.

—— 1984 [1967]: *The Magical Arts.* London: Arkana.

—— 1990: *A History of Magic.* London: Arkana.

Chapman, J. 1993: *Quest for Dion Fortune.* York Beach, ME: Samuel Weiser.

Chapman, M. 1992: *The Celts: The Construction of Myth.* New York: St Martin's Press.

Chodorow, N. 1978: *The Reproduction of Mothering.* Berkeley: University of California Press.

Clark, M. 1990: *Nietzsche on Truth and Philosophy.* Cambridge: Cambridge University Press.

Cohen, A. and N. Rapport, eds. 1995: *Questions of Consciousness.* London: Routledge.

Cohn, N. 1970: 'The Myth of Satan and His Human Servants', *in Witchcraft Confessions and Accusations*, ed. M. Douglas. London: Tavistock [ASA Vol.:9].

—— 1993 [1975]: *Europe's Inner Demons.* London: Pimlico.

Comaroff, J. 1986: *Body of Power, Spirit of Resistance.* Chicago: University of Chicago Press.

Connell, R. 1987: *Gender and Power: Society, the Person and Sexual Politics.* Oxford: Polity Press.

Connor, S. 1989: *Postmodernist Culture: An Introduction to Theories of the Contemporary.* Oxford: Blackwell.

Constantinides, P. 1985: 'Women Heal Women: Spirit Possession and Sexual Segregation in a Muslim Society', *Social Science and Medicine*, Vol. 21 no. 21: 685–92.

Cooper, J. 1981: *Taoism: The Way of the Mystic*. Wellingborough: Aquarian (Originally published in 1972).

Cornwall, A. and N. Lindisfarne, eds. 1993: *Dislocating Masculinities*. London: Routledge.

Crapanzano, V. 1980: *Tuhami: Portrait of a Moroccan*. Chicago: University of Chicago Press.

Crapanzano, V. and V. Garrison, eds. 1977: *Case Studies in Spirit Possession*. New York: John Wiley.

Crowley, A. 1979 [1922]: *Diary of a Drug Fiend*. London: Abacus.

—— 1991 [1976]: *Magick in Theory and Practice*. Secaucus, NJ: Castle. (Originally printed as *Magick by the Master Therion*. Paris, 1929.)

Crowley, V. 1989: *Wicca: The Old Religion in the New Age*. London: Aquarian.

—— 1993: 'Women and Power in Modern Paganism', in *Women as Teachers and Disciples in Traditional and New Religions*, ed. E. Puttick and P. Clarke. Lampeter: The Edwin Mellen Press.

D'Alviella, G. 1981: *The Mysteries of Eleusis: The Secret Rites and Rituals of the Classical Greek Mystery Tradition*. Wellingborough: Aquarian.

Daly, M. 1985 [1973]: *Beyond God the Father*. London: The Women's Press.

Davidson, J. and R. Davidson, eds. 1980: *The Psychobiology of Consciousness*. New York: Plenum Press.

de Beauvoir, S. 1973 [1949]: *The Second Sex*. Harmondsworth: Penguin.

Derrida, J. 1976: *Of Grammatology*. Baltimore, MD: Johns Hopkins University Press.

Douglas, M. 1966: *Purity and Danger: An Analysis of Concepts of Danger and Taboo*. London: Routledge and Kegan Paul.

—— ed. 1970: *Witchcraft: Confessions and Accusations*. London: Tavistock.

—— 1975: *Implicit Meanings: Essays in Anthropology*. London: Routledge and Kegan Paul.

Dowell, S. and J. Williams, eds. 1994: *Bread, Wine and Women: The Ordination Debate in the Church of England*. London: Virago.

Duerr, P. H. 1985: *Dreamtime: Concerning the Boundary Between Wilderness and Civilization*, trans. Felicitas Goodman. Oxford: Blackwell.

Dumont, L. 1985: 'A Modified View of Our Origins: The Christian Beginnings of Modern Individualism', in *The Category of the Person*, ed. M. Carrithers, S. Collins, and S. Lukes, pp. 92–121. Cambridge: Cambridge University Press.

Durkheim, E. 1965 [1915]: *The Elementary Forms of the Religious Life: A Study in Religious Sociology*, trans. by Joseph Swain. New York: The Free Press.

—— 1992 [1957], *Professional Ethics and Civil Morals*. London: Routledge.

Easton, D. and C. Liszt, 1997: 'Sex, Spirit & S/M'. *Green Egg*, May–June, Vol. 29 no. 119:16–19.

Edwards, A. 1989: 'The Sex/Gender Distinction: has it Outlived its Usefulness?' *Australian Feminist Studies*, Vol. 10.

Edwards, G. 1992: 'Does Psychotherapy Need a Soul?', in *Psychotherapy and Its Discontents*, ed. W. Dryden and C. Feltham, pp. 202–11. Milton Keynes: Open University Press.

Eilberg, H. 1989: 'Witches of the West: Neopaganism and Goddess as Enlightenment Religions', *Journal of Feminist Studies in Religion*, 5 no.1.

Eliade, M. 1989 [1954]: *The Myth of the Eternal Return: Cosmos and History.* London: Arkana.

Erickson, V. 1992: 'Back to the Basics: Feminist Social Theory, Durkheim and Religion', *Journal of Feminist Studies in Religion*, 8: No.1.

Evans-Pritchard, E. E. 1956: *Nuer Religion.* Oxford: Oxford University Press.

—— 1985 [1937]: *Witchcraft Oracles and Magic Among the Azande.* Oxford: Clarendon.

Evola, J. 1983: *The Metaphysics of Sex.* London: East West Publications.

Faivre, A. 1989: 'What is Occultism?', in *Hidden Truths: Magic, Alchemy and the Occult*, ed. Lawrence Sullivan, New York: Macmillan.

Farrar, S. and J. Farrar 1991 [1984]: *A Witches' Bible Compleat.* New York: Magickal Childe Publishing.

Favret-Saada, J. 1980: *Deadly Words: Witchcraft in the Bocage.* Cambridge: Cambridge University Press.

Feinstein, D. and S. Krippner 1989: *Personal Mythology.* London: Unwin Press.

Ferguson, M. 1982: *The Aquarian Conspiracy.* London: Paladin.

Flax, J. 1991: *Thinking Fragments: Psychoanalysis, Feminism and Postmodernism in the Contemporary West.* Berkeley, CA: University of California Press.

Flint, V. 1993: *The Rise of Magic in Early Medieval Europe.* Oxford: Clarendon.

Forsyth, N. 1989 [1987]: *The Old Enemy.* Princeton, NJ: Princeton University Press.

Fortune, D. 1957: *The Demon Lover.* London: Aquarian.

—— 1960: *The Goatfoot God.* London: Aquarian.

—— 1976 [1927]: *The Cosmic Doctrine.* Wellingborough: Aquarian.

—— 1985 [1956]: *Moon Magic.* York Beach, ME: Samuel Weiser.

—— 1987a: *Applied Magic and Aspects of Occultism.* Wellingborough: Aquarian.

—— 1987b [1935]: *The Mystical Qabalah.* London: Aquarian.

—— 1988: *Psychic Self-Defence. London*: Aquarian .

—— 1989 [1959]: *The Sea Priestess.* Wellingborough: Aquarian.

—— 1993: *The Magical Battle of Britain.* London: Golden Gates.

Foucault, M. 1990 [1976]: *The History of Sexuality,* Vol. 1. London: Penguin.

Frazer, J. 1993 [1921]: *The Golden Bough.* Ware, Herts: Wordsworth.

Friedman, M. 1992: 'Feminism and Modern Friendship: Dislocating the Community', in *Communitarianism and Individualism*, ed. S. Avineri and A. de-Shalit, pp. 101–19. Oxford: Oxford University Press.

Frosh, S. 1987: *The Politics of Psychoanalysis*. Basingstoke: Macmillan.

—— 1991, *Identity Crisis*. Basingstoke: Macmillan.

Gadon, E. 1990: *The Once and Future Goddess*. Wellingborough: Aquarian.

Gage, M. 1982 [1893]: *Woman, Church and State: A Historical Account of the Status of Woman through the Christian Ages*. New York: Arno Press.

Galian, L. and C. Rizzo 1995: 'Can You Spell Hierarchy?', *Pagan Times*, 12(3).

Gardner, G. 1988 [1954]: *Witchcraft Today*. New York: Magickal Childe.

Geertz, C. 1973: *The Interpretation of Cultures*. New York: Basic Books.

Gellner, E. 1985: *The Psychoanalytic Movement: Or the Cunning of Unreason*. London: Paladin.

—— 1992a: *Reason and Culture: The Historic Role of Rationality and Rationalism*. Oxford: Blackwell.

—— 1992b: *Postmodernism, Reason and Religion*. London: Routledge.

Gerth, H. and C. Wright Mills, eds. 1970 [1948]: *From Max Weber: Essays in Sociology*. London: Routledge & Kegan Paul.

Giddens, A. 1991: *Modernity and Self Identity*. Cambridge: Polity Press.

Gilligan, C. 1993 [1982]: *In a Different Voice*. Cambridge, MA: Harvard University Press.

Gimbutas, M. 1982: *The Goddesses and Gods of Old Europe: Myths and Cultural Images*. Berkeley, CA: University of California Press.

—— 1989: *The Language of the Goddess*. San Francisco: Harper & Row.

Ginzburg, C. 1992: *Ecstasies: Deciphering the Witches' Sabbath*. New York: Penguin.

—— 1993: 'Deciphering the Sabbath', in *Early Modern European Witchcraft*, ed. B. Ankarloo and G. Henningsen. Oxford: Clarendon Press.

Glock, C. and R. Stark, 1965: *Religion and Society in Tension*. Chicago: Rand McNally.

Godbeer, R. 1992: *The Devil's Dominion*. Cambridge: Syndicate.

Goddard, D. 1996: *The Sacred Magic of the Angels*, New York: Samuel Weiser.

Godwin, J. 1981: *Mystery Religions in the Ancient World*. London: Thames and Hudson.

Goodman, N. 1968: *Languages of Anthropology*. Indianapolis: Bobbs-Merrill. 1978: *Ways of Worldmaking*. Brighton: Harvester Press.

Goodrick-Clarke, N. 1990: *Paracelsus: Essential Readings*. Wellingborough: Aquarian.

—— 1992: *The Occult Roots of Nazism: Secret Aryan Cults and their Influence on Nazi Ideology*. London: I.B. Tauris.

Grant, K. 1976: *Cults of the Shadow*. New York: Samuel Weiser.

—— 1994 [1977]: *Nightside of Eden*. London: Skoob.

Graves, R. 1981 [1961]: *The White Goddess*. London: Faber & Faber.

Gray, W. 1989: *Between Good and Evil*. St. Paul, MN: Llewellyn.

—— 1992 [1975]: 'Patterns of Western Magic: A Psychological Appreciation', in *Transpersonal Psychologies*, ed. C. Tart. San Francisco: Harper.

Green, M. 1991: A *Witch Alone: Thirteen Moons to Master Natural Magic* London: Aquarian.

Greenwood, S. 1995a: "Current Research on Paganism and Witchcraft in Britain" (with G. Harvey, A. Simes and M. Nye), *Journal of Contemporary Religion*, May 1995, Vol. 10 (2).

—— 1995b: '"Wake the Flame Inside Us": Magic , Healing and the Enterprise Culture in Contemporary Britain', in *Etnofoor,* Amsterdam, Vol. 8 no. 1.

—— 1995c: 'Feminist Witchcraft: a Transformatory Politics,' *in Practising Feminism: Identity, Difference and Power*, ed. F. Hughes-Freeland and N. Charles. London: Routledge.

—— 1995d: 'The Magical Will, Gender and Power in Magical Practices', in *Paganism Today*, ed. G. Harvey and C. Hardman. London: Aquarian.

—— 1996: 'The British Occult Subculture: Beyond Good and Evil?', in *Magical Religion and Modern Witchcraft*, ed. J. Lewis. Albany, NY: State University of New York Press.

—— 1998: 'The Nature of the Goddess: Sexual Identities and Power in Contemporary Witchcraft', in *Nature Religion Today: Paganism in the Modern World*, ed. Joanne Pearson, Richard Roberts and Geoffrey Samuel. Edinburgh: Edinburgh University Press.

—— 2000: 'Power and Gender in Paganism', in *Beyond the New Age: Spirituality and Religion in the Twentieth Century*, ed. Steven Sutcliffe and Marion Bowman. Edinburgh: Edinburgh University Press.

—— Forthcoming: *Nature Religion: Paganism, New Age and Western Shamanism* (in preparation).

Grof, S. 1976: *Realms of the Human Unconscious*. New York: Viking Press.

—— 1988: *The Adventure of Self-Discovery: Dimensions of Consciousness and New Perspectives in Psychotherapy and Inner Exploration*. Albany: State University of New York Press.

Halevi, Z. 1979: *Kabbalah*. London: Thames and Hudson.

—— 1986: *Psychology and Kabbalah*. Bath: Gateway Books.

Harding, E. 1982 [1955]: *Woman's Mysteries*. London: Rider.

Harford, B. and S. Hopkins, eds. 1984: *Greenham Common: Women at the Wire*. London: The Women's Press.

Harner, Michael 1990 [1980]: *The Way of the Shaman*. San Francisco: Harper.

Harré, R. 1993: *Personal Being*. Oxford: Blackwells.

Harris, A. 1996: 'Sacred Ecology', in *Paganism Today*, ed. Graham Harvey and Charlotte Hardman. London: Thorsons.

Harrison, J. 1979: *The Second Coming*. London: Routledge & Kegan Paul.

Harvey, D. 1991: *The Condition of Postmodernity*. Oxford: Blackwell.

Harvey, G. 1993: 'Avalon From the Mists: The Contemporary Teaching of Goddess Spirituality', *Religion Today*, Vol. 8 no. 2:10–13.

—— 1994 'The Roots of Pagan Ecology', *Religion Today*, Vol. 9 no. 3:38–41.

—— 1996 'Heathenism: A North European Pagan Tradition' in *Paganism Today*, ed. Graham Harvey and Charlotte Hardman. London: Thorsons.

—— 1997: *Listening People, Speaking Earth*. London: Hurst & Co.

Haug, F. ed. 1987: *Female Sexualization*. London: Verso.

Heald, S. and A. Deluz 1994: *Anthropology and Psychoanalysis: An Encounter Through Culture*. London: Routledge.

Hebdige, D. 1979: *Subculture: The Meaning of Style*. London: Methuen.

Heelas, P. 1982: 'Californian Self Religions and Socializing the Subject', in *New Religious Movements: A Perpective for Understanding Society*, ed. E. Barker, pp. 69–85. New York: The Edwin Mellen Press.

—— 1987: 'Exegesis: Methods and Aims', in *The New Evangelists*, ed. P. Clarke, pp. 17–30. London: Ethnographica.

—— 1992: 'New Age Ethics and BCCI', *Religion Today*, Vol. 8 no. 1:1–4.

—— 1993: 'The New Age in Cultural Context: The Premodern, the Modern and the Postmodern', *Religion*, 23:103–16.

—— 1995: 'De-Traditionalization of Religion and Self: The New Age and Post-modernity', in *Postmodernity, Sociology and Religion*, ed. K. Flanagan and P. Jupp. London: Macmillan.

—— 1996: *The New Age Movement: The Celebration of the Self and the Sacraliza-tion of Modernity*. Oxford: Blackwell.

Heelas, P. and R. Kohn 1986: 'Psychotherapy and Techniques of Transformation', in *Beyond Therapy*, ed. G. Claxton, pp. 293–310.London: Wisdom Publications.

Henning, J. 1989: 'God the Lover: Women, the Goddess and the Young God," *Wood and Water*, Vol. 2 no. 28:8–9.

Hester, M. 1992: *Lewd Women and Wicked Witches: A Study of the Dynamics of Male Domination*. London: Routledge.

Hewitson-May, G. 1992: *Dark Doorway of the Beast*. Yorkshire: New World Publishing.

Hine, P. 1992: *Condensed Chaos*. London: Phoenix Publications.

Hobsbawn, E. and T. Ranger, eds. 1983: *The Invention of Tradition*. Cambridge: Cambridge University Press.

Hole, C. 1986: *Witchcraft in Britain*. London: Paladin.

Hutton, R. 1991: *The Pagan Religions of the Ancient British Isles*. London: BCA.

Bibliography

—— 1993: *The Shamans of Siberia*. Somerset: The Isle of Avalon Press.

—— 1994: 'Neo-Paganism, Paganism and Christianity', *Religion Today*, Vol. 9 no. 3: 29–32.

—— 1996: 'The Roots of Modern Paganism' in *Paganism Today*, ed. Graham Harvey and Charlotte Hardman, London: Thorsons.

Huxley, A 1996 [1945]: *The Perennial Philosophy*. London: Flamingo.

Inden, R. 1985: 'Hindu Evil as Unconquered Lower Self', in *The Anthropology of Evil*, ed. David Parkin. Oxford: Blackwell.

Jacques, M. and S. Hall 1983: *The Politics of Thatcherism*. London: Lawrence and Wishart.

Jones, A. 1996: *Dictionary of World Folklore*. Edinburgh: Larousse.

Jones, L., ed. 1987: *Keeping the Peace*. London: The Women's Press.

Jones, P. and N. Pennick 1995: *A History of Pagan Europe*. London: Routledge.

Jordanova, L. 1990 [1980]: 'Natural Facts: A Historical Perspective on Science and Sexuality', in *Nature, Culture and Gender*, ed. C. MacCormack and M. Strathern, pp. 42–69. Cambridge: Cambridge University Press. Judge, W. 1969 [1893]: *The Ocean of Theosophy*. Bombay: Theosophy Company India.

Jung, C. 1993 [1953]: *Psychology and Alchemy*. London: Routledge.

Kapferer, B. 1988: *Legends of People, Myths of State*. Washington, DC: Smithsonian.

—— 1991 [1983]: *A Celebration of Demons*. Oxford: Berg.

Keller, C. 1989: 'Feminism and the Ethic of Inseparability', in *Weaving the Visions*, ed. J. Plaskow and C. Christ, pp. 256–65. San Francisco: Harper and Row.

Kellner, D. 1992: 'Popular Culture and the Construction of Postmodern Identities', in *Modernity and Identity*, ed. J. Friedman and S. Lasch, pp. 141–75. Oxford: Blackwell.

Kelly, A. 1991: *Crafting the Art of Magic*, Book 1: *A History of Modern Witchcraft, 1939–1964*, Vol. 1. St. Paul, MN: Llewellyn Publications.

Kieckhefer, R. 1993: *Magic in the Middle Ages*. Cambridge: Cambridge University Press.

Kilbourne, B. and J. Richardson 1984 'Psychotherapy and New Religions', *American Psychologist*, Vol. 39 no. 3: 237–42.

King, F. 1975: Magic: *The Western Tradition*. London: Thames and Hudson.

—— 1990 [1970]: *Modern Ritual Magic, The Rise of Western Occultism*. Bridport, Dorset: Prism.

King, U. 1989: *Woman and Spirituality*. Basingstoke: Macmillan.

Kitto, H. D. F. 1968 [1951]: *The Greeks*. Harmondsworth: Penguin.

Klaniczay, G. 1990: *The Uses of Supernatural Power: The Transformation of Popular Religion in Medieval and Early-Modern Europe*. Cambridge: Polity Press.

Klibansky, R. and E. Panofsky, eds. 1964: *Saturn and Melancholy*. London: Belson.

Knight, C. 1991: *Blood Relations: Menstruation and the Origins of Culture*. New Haven, CT: Yale University Press.

Knight, G. 1978: *A History of White Magic*. Oxford: Mowbrays.

Kohn, R. 1991: 'Radical Subjectivity in "*Self Religions*" and the Problem of Authority', in *Religion in Australia,* ed. A. Black, pp. 133–50. Sydney: Allen & Unwin.

Kramer, H. and J. Sprenger 1948 [*c.*1494]: *The Malleus Maleficarum*. New York: Dover Publications.

Kuhn, T. 1970: *The Structure of Scientific Revolutions*. Chicago: University of Chicago Press.

Kvale, S. 1992: *Pyschology and Postmodernism*. London: Sage.

La Fontaine, J. 1977: 'The Power of Rights', *Man* n.s., Vol. 12 no. 3/4: 421–37.

—— 1985: *Initiation: Ritual Drama and Secret Knowledge Across the World*. Harmondsworth: Penguin.

—— 1994: *The Extent and Nature of Organised and Ritual Abuse: Research Findings*. London: HMSO.

Larner, C. 1981: *Enemies of God*. Oxford: Blackwell.

—— 1984: *Witchcraft and Religion*. Oxford: Blackwell.

Laughlin, C. 1994: 'Psychic Energy and Transpersonal Experience: A Biogenetic Structural Account of the Tibetan Dumo Yoga Practice', in *Being Changed*, ed. D. Young and J.-G. Goulet. Peterborough, Ontario: Broadview Press.

Leach, E. 1976: *Culture and Communication*. Cambridge: Cambridge University Press.

Leland, C. G. 1991 [1899]: *Aradia, or the Gospel of the Witches*. London.

Leland, S. and L. Caldecott, eds. 1983: *Reclaim the Earth*. London: The Women's Press.

LeMasters, C. 1989: 'Unhealthy Uniformities', *The Women's Review of Books* Vol. VII, no. 1:15.

Lévi, E. 1982 [1913]: *The History of Magic*. London: Rider.

Lévi-Strauss, C. 1968 [1962]: *The Savage Mind*. London: Weidenfeld and Nicolson.

—— 1979 [1963]: 'The Effectiveness of Symbols', in *Reader in Comparative Religion: An Anthropological Approach*, 4th edn, ed. W. Lessa and E. Vogt, pp. 318–27. New York: Harper & Row.

Lévy-Bruhl, L. 1966 [1910]: *How Natives Think*, trans. Lillian Clare. New York: Washington Square Press. (Originally published as *Les Fonctions mentales dans les sociétés inférieures.*)

Lewis, I. 1989 [1971]: *Ecstatic Religion*. London: Routledge.

Lienhardt, G. 1961: *Divinity and Experience*. Oxford: Oxford University Press.

Lindholm, C. 1990: *Charisma*. Oxford: Blackwell.

Lock, H. 1981: *Indigenous Psychologies*. London: Academic Press.

Löwith, K. 1967: 'The Philosophical Concepts of Good and Evil', in *Studies in Jungian Thought*, ed. C.G. Jung Institute. Evanston, IL: Northwestern University Press.

Luck, G. 1989: 'Theurgy and Forms of Worship in Neoplatonism', in *Religion, Science and Magic*, ed. J. Neusner, E. Frerichs, and P. V. McCracken Flesher. New York: Open University Press.

Lugh 1992 [1984]: *Old George Pickingill and the Roots of Modern Witchcraft*. Charlottesville, VA: Taray Publications.

Luhrmann, T. 1986: 'Witchcraft, Morality and Magic in Contemporary London', *International Journal of Moral and Social Studies*, Vol. 1 no. 1: 77–91.

—— 1989: *Persuasions of the Witch's Craft*. Oxford: Basil Blackwell.

MacCormack, C. and M. Strathern, eds. 1990 [1980]: *Nature Culture and Gender*. Cambridge: Cambridge University Press.

Macfarlane, A. 1970: *Witchcraft in Tudor and Stuart England: A Regional and Comparative Study*. New York: Harper and Row.

MacGregor-Mathers, S. L. 1976: *The Book of the Sacred Magic of Abra-Melin the Mage*. Wellingborough: Aquarian.

Macklin, J. 1977a: 'A Connecticut Yankee in Summer Land', in *Case Studies in Spirit Possession*, ed. V. Crapanzano and V. Garrison. New York: John Wiley.

—— 1977b: 'Psychocultural Exegesis of a Case of Spiritual Possession in Sri Lanka', in *Case Studies in Spirit Possession*, ed. V. Crapanzano and V. Garrison. New York: John Wiley.

Malinowski, B. 1954 [1948]: *Magic, Science, and Religion and Other Essays*. New York: Doubleday Anchor Books.

—— 1978 [1935]: *Coral Gardens and Their Magic*. New York: Dover Publications.

Marcus, G. and M. Fischer, 1986: *Anthropology as Cultural Critique: An Experimental Moment in the Human Sciences*. Chicago: University of Chicago Press.

Martin, D. and G. Fine, 1991: 'Satanic Cults, Satanic Play: Is "Dungeons and Dragons" a Breeding Ground for the Devil?' in *The Satanism Scare*, ed. James T. Richardson, Joel Best, and David Bromley. New York: Aldine de Gruyter.

Martin, E. 1990: 'Science and Women's Bodies', in *Body/Politic: Women and the Discourses of Science*, ed. M. Jacobus, pp. 69–81. London: Routledge.

—— 1993 [1987]: *The Woman in the Body: A Cultural Analysis of Reproduction*. Milton Keynes: Open University Press.

Marwick, M. 1964: 'Witchcraft as a Social Strain-Gauge'. *Australian Journal of Science*, Vol. 26: 263–8.

—— ed. 1990 [1970]: *Witchcraft and Sorcery*. London: Penguin Books.

Matthews, C. and J. Matthews 1985: *The Western Way*. London: Arkana.

Mauss, M. 1972 [1950]: *A General Theory of Magic*, trans. Robert Brain. London: Routledge & Kegan Paul.

McClintock, A. 1993: 'Maid to Order: Commercial S/M and Gender Power', in *Dirty Looks: Women, Pornography and Power*, ed. Pamela Church Gibson and Roma Gibson. London: British Film Institute.

McNay, L. 1992: *Foucault and Feminism*. Cambridge: Polity.

Melton, J. G. 1978: *The Encyclopaedia of American Religions*. Wilmington, NC: McGrath.

—— 1986: 'New Age Movement', in *Encyclopaedic Handbook of Cults in America*. New York: Garland.

—— 1993: 'Another Look at New Religions', *Annals, AAPSS* 527: 97–112.

Michelet, J. 1862: *La Sorcière*. Paris.

Moore, H. 1994: *A Passion for Difference: Essays in Anthropology and Gender*. Cambridge: Polity.

Morris, B. 1987: *Anthropological Studies of Religion*. Cambridge: Cambridge University Press.

—— 1991: *Western Conceptions of the Individual*. Oxford: Berg.

—— 1993 'Paracelsus: Magus and Medic?' *The Rosicrucian Beacon*, Winter: 32–4.

—— 1994a: *Anthropology of the Self: The Individual in Cultural Perspective*. London: Pluto Press.

—— 1994b: 'Matriliny and Mother Goddess Religion', *The Raven Anarchist Quarterly* 25.

—— 1996: *Ecology and Anarchism: Essays and Reviews on Contemporary Thought*. Malvern Wells: Images Publishing.

—— 1997: 'In Defence of Realism and Truth: Critical Reflections on the Anthropological Followers of Heidegger', in *Critique of Anthropology*, Vol. 17(3): 313–39.

Muchembled, R. 1993: 'Satanic Myths and Cultural Reality', in *Early Modern European Witchcraft*, ed. B. Ankarloo and G. Henningsen. Oxford: Clarendon Press.

Murray, K. 1989: 'The Construction of Identity in the Narratives of Romance and Comedy', in *Texts of Identity*, ed. J. Shotter and K. Gergen. London: Sage.

Murray, M. 1921: *The Witchcult in Western Europe: A Study in Anthropology*. Oxford: Clarendon Press.

Neal, M. 1993: 'Reflections on Western Magic', *Round Merlin's Table*, no. 83: 3–11.

Nelson, G. 1987: *Cults, New Religions and Religious Creativity*. London: Routledge & Kegan Paul.

Nietzsche, F. 1968: *The Will to Power*. London: Weidenfeld and Nicolson.

—— 1969: *Thus Spoke Zarathustra*. London: Penguin.

—— 1978: *Beyond Good and Evil*. Harmondsworth: Penguin.

Noll, R. 1985: 'Mental Imagery Cultivation as a Cultural Phenomenon: The Role of Visions in Shamanism', *Current Anthropology*, Vol. 26 no. 4.

—— 1996: *The Jung Cult: The Origins of a Charismatic Movement*. London: Fontana.

Oakley, A. 1982 [1972]: *Towards New Society: Sex, Gender and Society*. London: Temple Smith.

Obeyesekere, G. 1968: 'Theodicy, Sin and Salvation in a Sociology of Buddhism', in *Dialectic in Practical Religion*, ed. Edmund Leach. Cambridge: Cambridge University Press.

—— 1970: 'The Idiom of Demonic Possession: A Case Study', *Social Science and Medicine*, Vol. 4: 97–111.

—— 1977: 'Psychocultural Exegesis of a Case of Spirit Possession in Sri Lanka', in *Case Studies in Spirit Possession*, ed. V. Crapanzano and V. Garrison. New York: John Wiley.

—— 1981: *Medusa's Hair*. Chicago: Chicago University Press.

O'Connor and I. McDermott 1996: *Principles of NLP*. London: Thorsons.

O'Flaherty, W. 1976: *The Origins of Evil in Hindu Mythology*. Berkeley, CA: University of California Press.

Ortner, S. 1974: 'Is Female to Male as Nature is to Culture', in *Woman, Culture and Society*, ed. M. Rosaldo and L. Lamphere. Stanford, CA: Stanford University Press.

Overing, J. 1978: 'The Shaman as a Maker of Worlds: Nelson Goodman in the Amazon' in *MAN, The Journal of the Royal Anthropological Institute*. Vol. 25, pp. 602–19.

—— 1985: 'Introduction' to *Reason and Morality*. London: Tavistock.

Pagels, E. 1990 [1979]: *The Gnostic Gospels*. London: Penguin.

Palmer, S. 1994: *Moon Sisters, Krishna Mothers, Rajneesh Lovers: Women's Roles in New Religions*. New York: Syracuse University Press.

Parfitt, W. 1991: *The Elements of the Qabalah*. Shaftesbury, Dorset: Element.

Parkin, D. 1985: *The Anthropology of Evil*. Oxford: Blackwell.

Parrinder, G. 1982: *Avatar and Incarnation: A Comparison of Indian and Christian Beliefs*. New York: Oxford University Press.

Pengelly, J., Robert Hall, and Jem Dowse, 1997: *The Origins and History of the Pagan Federation, Including Excerpts from The Wiccan 1968–1981*. London: Pagan Federation.

Perera, S. 1981: *Descent to the Goddess*. Toronto: Inner City Books.

Pickering, W. 1984: *Durkheim's Sociology of Religion*. London: Routledge & Kegan Paul.

Plaskow, J. and C. Christ, ed. 1989: *Weaving the Visions: New Patterns in Feminist Spirituality*. San Francisco: Harper & Row.

Plumwood, V. 1993: *Feminism and the Mastery of Nature*. London: Routledge.

Pois, R. 1986: *National Socialism and the Religion of Nature*. London: Croom Helm.

Puttick, E. 1995: 'Sexuality, Gender and the Abuse of Power in the Master–Disciple Relationship: The Case of the Rajneesh Movement', *Journal of Contemporary Religion*, Vol.10 no.1: 29–40.

—— 1997: *Women in New Religions: In Search of Community, Sexuality and Spiritual Power*. Basingstoke: Macmillan.

Puttick E. and P. Clarke, eds. 1993: *Women As Teachers and Disciples in Traditional and New Religions.* Lampeter: The Edwin Mellen Press.

Rabinovitch, S. 1992: 'An ye harm none, do what ye will: neo-pagans and witches in Canada', Unpublished MA thesis. Department of Sociology, New York University.

Radford Ruether, R. 1985: *Womanguides: Readings Toward a Feminist Theology.* Boston: Beacon Press.

Ramazanoglu, C. 1989: *Feminism and the Contradictions of Oppression.* London: Routledge.

Ransom, J. 1938: *A Short History of the Theosophical Society 1875–1937.* Madras: Theosophical Publishing House.

Rees, K. 1996: 'The Tangled Skein: The Role of Myth in Paganism', in *Paganism Today*, ed. G. Harvey and C. Hardman. London: Aquarian.

Regardie, R. 1981 [1937]: *The Art of True Healing.* Jersey: SOL. (Private publication.)

—— 1986: *The Eye in the Triangle: An Interpretation of Aleister Crowley.* Phoenix, AZ: Falcon.

—— 1991a [1938]: *The Middle Pillar.* St. Paul, MN: Llewellyn Publications.

—— 1991b [1932], *The Tree of Life: A Study in Magic.* York Beach, ME: Samuel Weiser.

Reid, S. 1996: 'As I Do Will, So Mote It Be: Magic as Metaphor in Neo-pagan Witchcraft' in *Magical Religion and Modern Witchcraft*, ed. J. R. Lewis. Albany, NY: SUNY Press.

Richardson, A. 1987: *Priestess: The Life and Magic of Dion Fortune.* Wellingborough: Aquarian.

Richardson, J., J. Best, and D. Bromley, eds. 1991: *The Satanism Scare.* New York: Aldine de Gruyter.

Ricoeur, P. 1981: *Hermeneutics and the Human Sciences.* Cambridge: Cambridge University Press.

Ring, K. 1976: 'Mapping the Regions of Consciousness: A Conceptual Reformulation', *Journal of Transpersonal Psychology* 8(2):77–88.

Robbins, T. 1991 [1988]: *Cults, Converts and Charisma.* London: Sage.

'Robert' 1979: 'Gardnerian Witchcraft', *Quest* no. 38: 43–6.

Rohrlich, R. 1990: 'Prehistoric Puzzles', *The Women's Review of Books*, Vol. VII no. 9.

Roper, L. 1994: *Oedipus and the Devil.* London: Routledge.

Rosaldo, M. 1980: 'The Uses and Abuses of Anthropology: Reflections on Feminism and Cross-Cultural Understanding', *Signs*, vol. 5 no. 3: 389–417.

Rubin, G. 1975: 'The Traffic in Women: Notes on the "Political Economy" of Sex', in *Toward an Anthropology of Women*, ed. R. Reiter, pp. 157–210. New York: Monthly Review Press.

Samuel, G. 1990: *Mind, Body and Culture: Anthropology and the Biological*

Interface. Cambridge: Cambridge University Press.

—— 1993: *Civilized Shamans: Buddhism in Tibetan Societies.* Washington, DC: Smithsonian Institution Press.

—— 1997: 'The Vajrayana in the Context of Himalayan Folk Religion', in *Tibetan Studies*, ed. Helmut Krasser, Michael Torsten Much, Ernst Steinkellner and Helmut Tauscher. Vienna: Verlag der Österreichischen Akademie der Wissenschaften.

—— 1998: 'Paganism and Tibetan Buddhism: Contemporary Western Religions and the Question of Nature' in *Nature Religion Today*, ed. Joanne Pearson, Richard H. Roberts and Geoffrey Samuel. Edinburgh: Edinburgh University Press.

Sanders, A. 1984: *The Alex Sanders Lectures.* New York: Magickal Childe.

Sass, L. 1992: 'The Epic of Disbelief: The Postmodernist Turn in Contemporary Psychoanalysis', in *Psychology and Postmodernism*, ed. K. Steinar, pp. 166–82. London: Sage.

Schieffelin, E. 1985: 'Performance and the Cultural Construction of Reality', *American Ethnologist*, Vol. 12 no. 4: 707–23.

Schuman, M. 1980: 'The Psychophysiological Model of Meditation and Altered States of Consciousness', in *The Psychobiology of Consciousness*, ed. J. Davidson and R. Davidson. New York: Plenum Press.

Segal, R. 1992: *The Gnostic Jung.* London: Routledge.

Seidler, V. 1989: *Rediscovering Masculinity: Reason, Language and Sexuality.* London: Routledge.

Shallcrass, P. 1996: 'Druidry Today', in *Paganism Today*, ed. Graham Harvey and Charlotte Hardman. London: Thorsons.

Sharma, U. 1978: 'Theodicy and the Doctrine of Karma', in *Man's Religious Quest*, ed. W. Foy. London: Croom Helm.

Sharpe, J. 1997: *Instruments of Darkness: Witchcraft in England 1550–1750.* London: Penguin.

Sherwin, R. 1993: 'On the Other Hand Path', *Nuit Isis*, Autumn Equinox: 96–101.

Shilling, C. 1993: *The Body and Social Theory.* London: Sage.

Shor, R. 1959: 'Hypnosis and the Concept of the Generalized Reality-Orientations', *American Journal of Psychotherapy*, Vol. 13: 582–602.

Shotter, J. and K. Gergen, eds. 1989: *Texts of Identity.* London: Sage.

Shweder, R. 1991: *Thinking Through Cultures: Expeditions in Cultural Psychology.* Cambridge, MA: Harvard University Press.

Singer, J. 1989: *Androgyny: The Opposites Within.* Boston: Sigo Press.

Skorupski, J. 1976: *Symbol and Theory*, Cambridge: Cambridge University Press.

Southwold, M. 1985: 'Buddhism and Evil', in *The Anthropology of Evil*, ed. D. Parkin. Oxford: Blackwell.

Starhawk 1982: *Dreaming the Dark: Magic, Sex and Politics.* Boston: Beacon.

—— 1989 [1979]: *The Spiral Dance*. San Francisco: Harper & Row.

—— 1990: *Truth or Dare: Encounters with Power, Authority, and Mystery*. San Francisco: Harper & Row.

Stewart, R. 1991: *Celebrating the Male Mysteries*. Bath: Arcania.

Stoller, P. (and C. Olkes) 1987: *In Sorcery's Shadow*. Chicago: University of Chicago Press.

Stone, M. 1976: *The Paradise Papers: The Suppression of Women's Rites*. London: Virago.

Symonds, J. and K. Grant 1989: *The Confessions of Aleister Crowley*. London: Arkana.

Tambiah, S. 1979: 'The Form and Meaning of Magical Acts: A Point of View', in William Lessa and Evon Vogt (eds), *Reader in Comparative Religion*. New York: Harper & Row.

—— 1985 [1970]: *Buddhism and the Spirit Cults in North-East Thailand*. Cambridge: Cambridge University Press.

—— 1991: *Magic, Science, Religion, and the Scope of Rationality*. Cambridge: Cambridge University Press.

Tarn, W. and G. T. Griffith 1952: *Hellenistic Civilisation*. London: Edward Arnold.

Tart, C. 1969: *Altered States of Consciousness*. New York: Wiley.

—— 1980: 'A Systems Approach to Altered States of Consciousness', in *The Psychobiology of Consciousness*, ed. J. Davidson and R. Davidson. New York: Plenum Press.

—— 1992 [1975]: *Transpersonal Psychologies*. San Francisco: Harper.

Taussig, M. 1991 [1986]: *Shamanism, Colonialism and the Wild Man*. Chicago: The University of Chicago Press.

Taylor, C. 1989: *Sources of the Self*. Cambridge: Cambridge University Press.

Tedlock, B. 1991: 'From Participant Observation to the Observation of Participation: The Emergence of Narrative Ethnography', *Journal of Anthropological Research*, 47: 69–94.

Thomas, K. 1971: *Religion and the Decline of Magic: Studies in Popular Beliefs in Sixteenth- and Seventeenth-Century England*. London: Penguin.

Thompson, K. 1992: 'Individual and Community in Religious Critiques of the Enterprise Culture', in *The Values of the Enterprise Culture*, ed. P. Heelas and P. Morris, pp. 253–75. London: Routledge.

Tipton, S. 1982: *Getting Saved from the Sixties*. London: University of California Press.

Trevor-Roper, H. R. 1967: *Religion, the Reformation and Other Essays*. London: Penguin.

Turner, E. 1992: *Experiencing Ritual: A New Interpretation of African Healing*. (with William Blodgett, Singleton Kahona and Fideli Benwa). Philadelphia: University of Pennsylvania Press.

—— 1994: 'A Visible Spirit Form in Zambia' in *Being Changed by Cross-Cultural Encounters: The Anthropology of Extraordinary Experience*, ed. David E. Young and Jean-Guy Goulet. Peterborough, Ontario: Broadview Press.

Turner, V. 1967: *The Forest of Symbols: Aspects of Ndembu Ritual*. Ithaca, NY: Cornell University Press.

—— 1974 [1969]: *The Ritual Process: Structure and Anti-Structure*. Harmondsworth: Penguin.

—— 1974: *Dramas, Fields and Metaphors*. London: Cornell University Press.

Tylor, E. 1871: *Primitive Culture*. London: John Murray.

Valiente, Doreen. 1978: *Witchcraft for Tomorrow*. Washington, DC: Phoenix.

—— 1986 [1973]: *An ABC of Witchcraft: Past and Present*. London: Hale.

Van Dijk, R. and P. Pels (forthcoming): 'Contested Authorities and the Politics of Perception: Deconstructing the Study of Religion in Africa', in *Postcolonial Identity in Africa*, ed. T. Ranger and R. Werbner. London: Zed Books.

Van Gennep, A. 1961 [1908]: *Rites of Passage*. Chicago: University of Chicago Press.

Versluis, A. 1986: *The Philosophy of Magic*. Boston: Arkana.

Vesey, G. and P. Foulkes, 1990: *Dictionary of Philosophy*. London: Collins.

Walker, B. 1982: *Tantrism*. Wellingborough: Aquarian.

Walsh 1990: *The Spirit of Shamanism*. London: Mandala.

Wang, R. 1990: *The Qabalistic Tarot*. York Beach, ME: Samuel Weiser.

Washington, P. 1993: *Madame Blavatsky's Baboon*. London: Secker and Warburg.

Watts, A. 1981: *Tao: The Watercourse Way*. Harmondsworth: Penguin.

Webb, J. 1976: *The Occult Establishment*. La Salle, IL: Open Court.

Weber, M. 1966 [1922]: *The Sociology of Religion*. London: Methuen.

—— 1978 [1930]: *The Protestant Ethic and the Spirit of Capitalism*. London: Allen & Unwin.

Wehr, D. 1989: *Jung and Feminism*. London: Routledge.

Wehr, G. 1990: *The Mystical Marriage*. Wellingborough: Aquarian.

Whitmont, E. 1987: *Return of the Goddess*. London: Arkana.

Wise, C. 1989: 'Book Review: *Wicca: The Old Religion in the New Age*, by Vivianne Crowley', *The Ley Hunter*, 18.

Wright, E., ed. 1992: *Feminism and Pyschoanalysis*. Oxford: Blackwell.

Yates, F. 1986 [1972]: *The Rosicrucian Enlightenment*. London: ARK.

—— 1991 [1964]: *Giordano Bruno and the Hermetic Tradition*. Chicago: University of Chicago Press.

York, M. 1994: 'New Age in Britain', *Religion Today*, Vol. 9 no. 3:14–21.

—— 1995: *The Emerging Network: A Sociology of the New Age and Neo-Pagan Movements*. Lanham, MD: Rowman & Littlefield.

Young, A. 1990: *Femininity in Dissent*. London: Routledge.

Bibliography

Young, D. 1994: 'Visitors in the Night: A Creative Energy Model of Spontaneous Visions', in *Being Changed*, ed. D. Young and J.-G. Goulet. Peterborough, Ontario: Broadview Press.

Young, D. and J.-G. Goulet, eds. 1994: *Being Changed By Cross-Cultural Encounters: The Anthropology of Extraordinary Experience.* Peterborough, Ontario: Broadview Press.

Young, M. 1990: 'The Ideal of Community and the Politics of Difference', in *Feminism/Postmodernism*, ed. L. Nicholson, pp. 300–23. New York: Routledge.

Index

re-enchantment and, 121
religious aspect of, 23, 179, 210
science and, 13
sympathetic 38
the will and, 37, 39, 41, 57, 124, 132–3, 144,
    152–3, 171–5, 197, 207
training and, 26, 33, 49–82, 121
wisdom and, 189
women's and men's, 167–70
magical ideologies, 1, 111–14, 119, 138, 142,
    147, 151, 159, 165, 175, 202, 211
magical power, 23
    abuse of, 174–5, 200–2, 208, 211
    chants and, 124
    childhood abuse and, 138–9, 142, 148n6
    crones, hags and, 31–2, 87
    forces and energies, 23, 39, 53–4, 65, 67,
        77–8, 89, 121, 132–7, 161, 179, 209
    good, evil and, 179
    meditation and, 79
    nature and, 190
    psyche and, 136–9, 163, 202
    questioning of, 141
    reclamation of, 130, 152
    ritual as channel of, 30, 34, 38, 87–8
    sacred space and, 41, 46n9, 58–9, 144–7
    sex and, 137, 155, 175–77, 200
    women and, 103–4, 138, 151–2, 159, 165,
        168–9, 176–7
    wyrd and, 199
    see also otherworld
    see also charisma
magical subculture in London 3, 18
    New Age and, 9
magical transformation, 2, 10, 95
    ecological action and, 111–12
    hierarchy and, 144–5
    initiation and, 30–1
    power and, 176–7
    psychotherapy and, 89
    self and, 35–7, 80, 142–3
    story-telling and, 28
    see also invocation
Malinowski, Bronislaw, 38–9, 47n11
masculinity, 1, 68, 70, 77, 94, 103–4, 151–8,
    161–4, 169, 171–2
    see also gender
Mauss, Marcel, 38–9

meditation, 3, 209
    exercises and practice, 16, 61–8
    guided visualization, 29, 61, 170, 192–3
    Kabbalah and, 68–71, 156–7
    on tarot, 52–60
    overcoming abuse and, 142
    see also pathworking
methodology
    see reflexive anthropology
Michelet, Jules, 196–7
microcosm/macrocosm, 23, 29, 77, 174, 206
    body and, 31, 36–7, 85
    duality and, 193
    early Greek world view and, 198
    healing and, 117, 124–5, 132
    magical self and, 145
    spiritual evolution and, 156
Moore, Henrietta, 119
morality, 1, 2, 57
    abuse of power and, 174–5, 200
    good, evil and, 2, 37–9, 58
    internal, 3, 137, 184–205
    moral discourse, 182–4
    see also Karma
Morris, Brian, 43–4, 132–3, 198
Murray, Kevin, 118–9, 147n1
Murray, Margaret, 83, 103
mysticism, 78, 125
mythology, 62, 209, 212
    gender and, 173
    identity and, 120
    Judgement of Osiris and, 184–6
    of Aradia, 196
    of Atlantis, 154, 155, 194
    of Demeter and Persephone, 66–7
    of Inanna, 31, 60, 128, 131, 157–8, 170
    otherworld and, 24, 27, 29, 61, 65
    personal myths and, 119
    ressurection and, 158

nature religion, 5, 83–9, 91, 109, 174, 203,
    111–13
    Earth Healing Day and, 122–4
    environmental movements and, 135
    folk magic and, 7
    see also hedgewitches
    witchcraft as, 7–8
Nazism, 193

Index